The Beginner's Guide to Amateur Radio

The Beginner's Guide to Amateur Radio

By the Staff of
THE AMERICAN RADIO
RELAY LEAGUE, INC.

PRENTICE-HALL, INC., Englewood Cliffs, NJ 07632

Library of Congress Cataloging in Publication Data
Main entry under title:

The Beginner's guide to amateur radio.

Includes index.
1. Radio—Amateurs' manuals. I. American Radio Relay League.
TK9956.B39 621.3841'66 81-7364
ISBN 0-13-072157-3 AACR2
ISBN 0-13-072140-9 (pbk.)

Editorial/production supervision
and interior design by *Daniela Lodes*
Cover illustration courtesy of *The American Radio Relay League*
Manufacturing buyer: *Gordon Osbourne*

Printed in the United States of America
10 9 8 7 6 5 4 3 2

Prentice-Hall International, Inc., *London*
Prentice-Hall of Australia Pty. Limited, *Sydney*
Prentice-Hall of Canada, Ltd., *Toronto*
Prentice-Hall of India Private Limited, *New Delhi*
Prentice-Hall of Japan, Inc., *Tokyo*
Prentice-Hall of Southeast Asia Pte. Ltd., *Singapore*
Whitehall Books Limited, Wellington, *New Zealand*

CONTENTS

Preface

When a European Space Agency rocket hurls a small satellite into earth orbit sometime during 1982, amateur radio operators the world over will rejoice—still another means of communication will have been made available to them. The launch of this satellite, will bring the entire Northern Hemisphere within range of amateur radio signals.

From the earliest days of radio, when Marconi labored for years to span the Atlantic via radio waves, amateur radio operators have been at the forefront of communications technology. Yet it doesn't take a great deal of technical skill to qualify for an amateur radio license. Despite differences in language, national origin and political systems, the one million amateurs worldwide are able to cut across artificial boundaries to make friends the world over—right from their armchairs. No other hobby can make that claim!

This book provides everything the beginner should know to prepare to enter a world that most people just dream about. It covers all areas of this fascinating pastime—setting up a station, building a simple wire antenna, getting into a licensing class that will end with an amateur radio license, getting involved with the many specialized activities, such as satellite communications and DXing (communicating over long distances).

Written by members of the headquarters staff of the American Radio Relay League, the U.S. and Canadian organization "of, by and for the amateur," this book is packed with easy-to-understand information that will provide the groundwork for entering the exciting, ever-expanding world of amateur radio. For more information, write Amateur Radio, Newington, Connecticut 06111.

Introduction

Tune your multiband ("shortwave") radio to the 1.6-megahertz band and what do you find? Around 1.650 you may hear a police cruiser reporting in to his dispatcher or asking assistance on a robbery call. At about 2.100 you could hear a ship's operator reporting his position to the Coast Guard or notifying other ships of a just-spotted iceberg hazard. At other spots on the dial you can pick up weather reports and conversations between aircraft crews and control towers. All these different radio contacts use the human voice—recognizable if not always understandable. At about 3.700 you may come across something different—the rapid dits and dahs of Morse code signals. Ah, you say, what's this?

"This" is amateur radio, perhaps better known as ham radio, a very special kind of hobby and a very special fraternity. Those Morse code signals are coming from hams, in fact probably from newly licensed hams called Novices, because 3.700 to 3.750 is one of the bands assigned specifically to their use. Novices use just Morse code for their on-the-air conversations, but hams with higher-class licenses "talk" on other amateur bands using voice, radioteletype, and even TV.

That is one of the differences, in fact, between ham radio and citizens' band radio, with which it is sometimes confused. CBers can use only voice. Hams can, and do, talk around the world. CBers are limited by law to 150 miles. Hams can use up to 1000 watts of power; CBers are limited to 5. Hams have access to many different frequencies in several bands; CBers can use just 40 specific channels in one band. In other words, hams can communicate on a much broader,

FIG. 1-1 Tom and Mary Moore, who hold the call signs KL7Q and KL7P, exemplify the horizons open to amateur radio operators. Not only did they relay 2600 messages in their native state of Alaska on a recent Mother's Day, they are active all year round in many types of public service. Tom was communications chairman of a 2-week dogsled race from Anchorage to Nome as well as the State Trooper's Run for the Special Olympics. Mary is membership chairman of her local amateur radio club and has served on many of the same communications projects as her husband. Studying together, they have earned the highest type of amateur license, the Extra class. (Photo by Glenn Alvord, KL7HX.)

more liberal basis, and over unlimited distances. For these greater operating privileges, hams must pass examinations on radio theory, rules and regulations, and Morse code in order to get their licenses, whereas CBers are licensed without examination.

If CB had done no more than to bring two-way communications to motorists and truckers on the nation's highways, it would still be one of the great developments of our time. For getting information en route, for assistance in local emergencies, for local business communications, or just to relieve the monotony of long trips, it is a useful and worthwhile radio service. For millions of Americans it has also been an intriguing introduction to personal radio communication. For thousands it has been the first step toward the wide world of amateur radio.

A Round-the-World Adventure

Ham radio, on the other hand, offers all of the mobile communications advantages of CB with a lot less congestion and clutter. But more than that, it means that with the flick of a switch, a twist of a dial, you are in instant personal touch with an accountant in London, a farmer in Kenya, a shopkeeper in Kyoto, an engineer in Rio de Janeiro, a rancher near Darwin, a doctor in Managua, or even a king in Amman! The day-to-day, people-to-people aspect of amateur radio, without limitations of geography, ideology, or even language, is perhaps its greatest and unique appeal.

When a freak tornado ripped through a quiet suburban neighborhood in Windsor, Connecticut, volunteer amateur radio operators from a 50-mile radius poured into the area. By providing the very type of emergency communications most needed by strapped local police and fire departments, the amateurs made the aftermath a great deal easier. The hams operated hand-held 2-meter equipment that enabled them to communicate with other amateurs, and disaster-relief agencies, in surrounding towns. With power and telephone lines down, amateur radio proved to be the most reliable form of emergency communication. The amateurs passed hundreds of "health and welfare" messages to worried friends and relatives of the tornado victims. In gratitude, the town's mayor praised the volunteers and asked area employers to give them time off from work so they could continue their vital mission. After a week, their help no longer needed, the scores of amateurs returned to their respective communities—until they are needed again.

Amateurs provide this type of community service at no cost; in fact, it is against the Federal Communications Commission (FCC) rules and regulations for a ham to accept any type of compensation for operating a station.

During a blizzard that hit the northern United States in January 1978, a young Michigan couple found themselves totally isolated and without telephone service when the imminent arrival of a new child became evident. Fortunately, their next-door neighbor, a ham, was able to contact their doctor and relay instructions using his car's amateur equipment. Mother and daughter both did nicely, but it took a couple of days for Dad to recover completely.

FIG. 1-2 When a disastrous earthquake struck Guatemala in 1976, amateur radio operators were the first to get word of the emergency to the outside world. A group of volunteers were the principal means of communication for days, serving government and relief agencies. (UPI photo.)

These are just two examples of the kinds of emergency communications services radio amateurs have been providing for more than a half-century. They range from major international disasters to local emergencies. This is a form of public service in which all hams take pride and satisfaction.

In hundreds of classrooms across the country, students have been getting firsthand exposure to live space communication through a satellite called OSCAR (for Orbiting Satellite Carrying Amateur Radio). In a program coordinated by the American Radio Relay League (ARRL), the national ham association, local operators take portable station equipment to schools and put on actual demonstrations using prearranged satellite contacts with ham stations in the United States and more than 80 other countries. The first OSCAR was built by a group of California amateurs in 1961—for out-of-pocket expenses of just $62! A series of increasingly sophisticated satellites has followed—all designed and built by hams and launched as hitchhikers with other satellites by NASA. Thousands of amateurs worldwide communicate via OSCAR and bring a taste of the space age into their homes.

Amateur radio is by no means all public service—satisfying as that aspect of their hobby is to hams. Ham radio is, after all, an exciting, diversified hobby, and today's hams enjoy it in many different ways. There are the DXers who delight in making contact with other amateurs in as many different countries as possible—especially

FIG. 1-3 Junior high school students get a live introduction to the space age via OSCAR, an orbiting communications satellite built by and for amateurs.

remote and isolated places. There are the contesters whose passion is to compete in the dozens of ham contests staged year-round—the purpose being to make as many contacts as possible with as many places as possible in a certain time period. There are the experimenters and builders to whom getting on the air is distinctly secondary. And then there are the great majority of all hams, who just enjoy getting on the air for a chat with a friend across the world or across town.

There are more than a million licensed radio amateurs around the world, most of them in Japan and the United States. There are more amateur stations in the United States than there are commercial broadcasters and government and military stations combined. Hams come from all walks of life, all kinds of backgrounds, and all ages from 8 to 80. The ham ranks include businesspeople, mechanics, farmers, shopkeepers, housewives, students, engineers, doctors—and just about any occupation you can name. They also include statesmen, professional athletes, show business personalities, and kings. Amateur radio comes about as close to pure democracy as you can get. A schoolboy may talk with Barry, K7UGA (Senator Goldwater); Hussein, JY1 (King of Jordan); Owen, W5LFL (NASA astronaut Owen Garriott); or Ronnie, WB4KCG (top country-western singer Ronnie Milsap). Amateur radio is always on a first-name basis—social or class distinctions are nonexistent. The common bond is a "ham ticket," the license that represents membership in one of the world's elite fraternities.

Amateur radio is literally as old as radio itself. Marconi, the father of wireless, considered himself an amateur, and many of the early radio pioneers were amateur operators. Amateurs have made major contributions to electronics technology as we know it today. The practical uses of the shortwave bands were discovered and developed by amateurs shortly after the turn of the century. Then thought to be useless, these frequencies had been allocated to amateurs, leaving the "good" frequencies for broadcast, government, and military use. Much of our present knowledge of radio-wave propagation—radio-wave behavior in the upper atmosphere, which determines when and how signals can be received—can be attributed to hams. And many of today's electronics engineers had their career interest first awakened by a youthful experience as amateurs.

People-to-People Communications

At the 14th World Scout Jamboree in Lillehammer, Norway, boys and girls—all Scouts—from more than 90 countries had the opportunity to participate in an active amateur radio program and demonstration. Amateur station LC1J was set up and put on the air by Scouts from the host Scandinavian countries; in less than a fortnight more than 2000 contacts were made with hams in 105 countries, using voice, Morse code, radioteletype, and slow-scan TV.

"UA2AB in Leningrad, Russia—this is VE6XYZ in Calgary. Your signal is loud and clear. Very pleased to meet you, Sergei. My name is Bill—Bravo India Lima Lima. Our weather is warm and sunny."

Not long ago, an amateur at Canadian Forces Base Alert, located 450 miles from the North Pole, made contact with the Admiral Byrd Station, just 1000 yards from the South Pole, for the longest recorded direct north-south communication ever.

"VE4YZ in Winnipeg, this is VO2ABC in Labrador. Bob, I have a brother in Winnipeg. I wonder if you would mind passing a personal message to him for me."

An off-duty U.S. Air Force sergeant got surprisingly fast results when he griped in an on-the-air conversation about poor delivery of needed parts to his remote base. He may have wondered how it came to pass that the parts were delivered the next day. He never realized that Curt, his ham contact, was then-Commanding General Curtis LeMay, W6EZV.

At the height of one of the Middle Eastern conflicts, King Hussein of Jordan, JY1, visited the United States and was a weekend house guest of Madeline (W2EEO) and Arthur (W2LH) Greenberg of New York. There are no international or ethnic tensions in amateur radio.

Dr. Greta Hubacher, HB9ARC, a dental surgeon in Winterthur, Switzerland, treasures this experience: "Ham radio gives, apart from the pleasure we can have ourselves and give to others, opportunity to help when it is needed. I had the opportunity to help supply a patient in Venezuela with a rare remedy that I could organize from here. This could only be got from Serum-Und Impf-Institut in Berne. Lots of hams joined me to help that this remedy could be brought under police supervision—for they needed empty roads—very fast to Zurich-Kloten airport and from there to Venezuela."

Austine Henry, VK3YL, of Victoria, Australia, was one of that continent's first women hams. She served before World War II in the Royal Australian Air Force Wireless Reserve, and during the war taught Morse code to servicemen. She caught the radio bug while convalescing from a childhood illness. As she tells the story, "A kindly uncle gave me a crystal set to cheer me up. I was entranced. Later, after much trial and error, I built a two-valve receiver (sorry, two tubes), which in turn led to short wave and the Morse code. After that it was only natural to want to transmit and communicate with other hams. I can still remember the excitement of grinding my own crystals from quartz spectacle lenses."

FIG. 1-4 The roster of those calling themselves hams includes rich and poor, celebrity and student, African farmer, and Canadian Mountie. Austine Henry, VK3YL, was first licensed in 1930 after being introduced to a small receiving set. Still going strong, this Australian amateur is active on the amateur satellites.

Barbara Duarte, PY1DVQ, of Rio de Janeiro, Brazil, tells how she taught Morse code to eight boys eager for ham licenses. "Twice a week they came at night to my house. We took places around my dining table that is 2 meters long and I began singing to them dits and dahs. Then I went to the key and had them recognizing all the alphabet. Then I started with words, but soon I found out that they were just guessing the words, so I changed from Portuguese to English. As most of them did not know English they had to identify the right signal. Later I started with phrases such as 'Wagner is not paying attention.' They learned very well."

From rural Wisconsin a former CBer writes: "Up until about a year ago I was very heavy into CB radio. A fellow I work with had some old radio gear in his attic which he wanted to sell. So I bought the whole works for $50. It included a transceiver, a power supply, an oscilloscope and some other stuff. After just listening in on the world of ham radio I was convinced I had to try for my first amateur ticket. I sent for a code tape and practiced everywhere I had the chance. I built a low-power transceiver from a kit, along with its power supply,

FIG. 1-5 Although the equipment may be complex, amateur radio's "bottom line" is quite simple—people talking to people. A young woman attending an exhibit at the Boston Museum of Science seems eager to grasp the basics of slow-scan television, just one of the many types of communication open to amateurs. (Photo by Joel Kleinman, WA1ZUY.)

and have found it can pull in some of the most enjoyable listening I have ever experienced."

Most people would consider communicating via radio signals reflected from the moon a pretty exotic activity. But not only have amateurs accomplished this feat, one has even achieved verified "moonbounce" contacts with hams on all the world's continents. In ham terms, he had earned the Worked All Continents (WAC) award on EME (earth–moon–earth). He is Dr. Allen Katz, K2UYH, chairman of the Engineering Technology Department at Trenton State College. Allen was first licensed as a radio amateur at the age of 12.

A Florida TV station and the *Miami Herald* reported the story of Jeff Wallis, WB4LGI, a blind ham. For many years he operated a candy and cigarette stand in the Miami Federal Building, but was forced to close it down after one too many robberies. He was offered a government job as a phone dispatcher but the job required him to use a five-line phone. Unfortunately, there was no way for him to identify which phone was ringing. A commercially available device was suggested, but the $1000 price tag was prohibitive. When fellow hams heard about his problem, one designed and another built a device with about $10 worth of solenoids which enables Jeff to identify the correct line by his sense of touch.

The Federal Communications Commission is the U.S. government agency that issues licenses and administers the Amateur Radio Service. One of its best known commissioners, first appointed by President Eisenhower, was Robert E. Lee. Commissioner Lee has summed up the appeal of amateur radio very well indeed: "Amateur radio is a consuming avocation in one's youth; often an indispensable foundation for a professional career in electronics and a source of tremendous pleasure and interest later in life."

And, as any ham will tell you, it is also a lot of fun.

2

Amateur Radio: Just What Is It?

Why, hams are often asked, is Morse code still a requirement for an amateur license? After all, with today's modern technology, voice is so much simpler and the code seems so old-fashioned.

The answer is that the international treaty agreements under which amateur radio operates require proficiency in the use of Morse code. Why? For one thing, code is still the surest way of getting a signal through under all kinds of conditions with the simplest equipment. That is why some of the military and other services which had switched to voice as their standard form of radio communication have now gone back to the code. Another reason is that the code is also both the simplest and most economical mode for universal, worldwide use. That is particularly important in the emerging nations of the Third World.

Some prospective hams, it is true, have seen the code require-

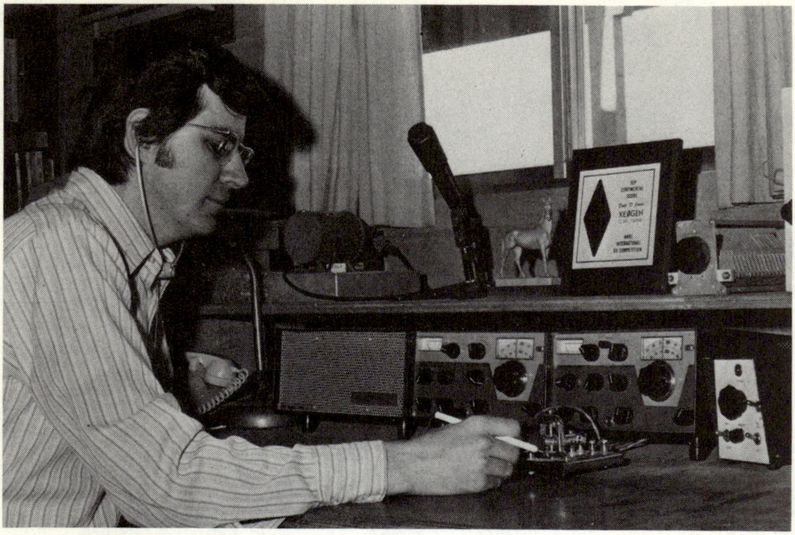

FIG. 2-1 Hams the world over use Morse code, the universal language that crosses cultural barriers with ease.

ment as a barrier that represents mastery of a skill which they have no particular interest in using. But as many new hams will happily testify, neither objection is really valid. The code is not difficult to learn, even for youngsters. And it is fun. It becomes a special language which makes its practitioners members of a special fraternity. Some say it is like learning another language, but it is in fact much simpler than that because it is just a set of new symbols for a familiar language. Even the process of learning the code is fun these days—either in a licensing class, with the help of another ham, or by using one of the many code tapes available.* With all the ways to communicate available to modern amateurs, code is still the first choice of many hams and is used on occasion by almost all hams.

What makes amateur radio so exciting, so satisfying as a lifelong pursuit, is its infinite variety, its unpredictability. It is a continuing process of discovery. Above all, amateur radio is for every ham a

*Write the American Radio Relay League, 225 Main Street, Newington, Connecticut 06111, for information.

very personal experience. That experience can have many dimensions.

To understand all the forms that experience can take, it is useful to start with the official view. The Amateur Radio Service is defined in the international regulations as:

> A service of self-training, inter-communication and technical investigations carried out by amateurs, that is by duly authorized persons interested in radio technique solely with a personal aim and without pecuniary interest.

The Federal Communications Commission, the licensing and regulatory agency for amateur radio in the United States, sets forth the purpose of the service in its regulations as follows:

97.1 Basis and Purpose.
The rules and regulations in this part are designed to provide an amateur radio service having a fundamental purpose as expressed in the following principles:

a. Recognition and enhancement of the value of the amateur service to the public as a voluntary, noncommercial communication service, particularly with respect to providing emergency communications.
b. Continuation and extension of the amateur's proven ability to contribute to the advancement of the radio art.
c. Encouragement and improvement of the amateur radio service through rules which provide for advancing skills in both the communication and technical phases of the art.
d. Expansion of the existing reservoir within the amateur radio service of trained operators, technicians and electronics experts.
e. Continuation and extension of the amateur's unique ability to enhance international good will.

Herbert Hoover, Jr., W6ZH, son of the former president and himself a former president of the American Radio Relay League, the national ham organization, stated it somewhat more briefly: "It was the determination that amateurs should have access to a limited but adequate share of the radio spectrum, under rules and regulations that would provide an incentive for voluntary public service, that originally made this activity possible—and has kept it going ever since."

A Resource Waiting to Be Mined

The radio spectrum to which Mr. Hoover referred is above all a natural resource—a finite natural resource. It is a resource much coveted by many powerful communications interests—radio and television broadcasters, national and international government agencies, the military, the maritime service, and the aeronautical industry, among others. Radio communication is absolutely necessary to the existence and operations of all of these. For an "amateur" body to be allocated substantial segments of the spectrum is evidence of the long and solid tradition of public service established by amateur radio operators—an activity that has won virtually universal recognition by the world's governments.

Amateur radio operators can be found in most countries of the world. Japan leads with about 700,000 amateurs, and the United States has nearly 400,000. There are more than 30,000 licensed hams in the Soviet Union, and the United Kingdom and West Germany each have more than 15,000 amateurs. Italy, Australia, Canada, Brazil, Argentina, Mexico, India, Sweden, Norway, and Israel all have large and active amateur populations. More than 100 national amateur societies, in fact, are members of the International Amateur Radio Union.

First recognized formally at the International Telecommunication Conference in Washington, D.C., in 1927, the Amateur Radio Service grew slowly until World War II. Since little or no manufactured equipment was marketed for amateur use, hams of necessity built their own station equipment. This ensured a high level of technical knowledge in the amateur ranks. Hams had a strong common bond and universally shared interests. Since World War II, however, amateur radio has changed a great deal. Radio technology accelerated during the war, as it did in so many other fields. And

FIG. 2-2 A typical amateur station of the 1930s boasted both homemade and commercial equipment. Despite an explosion in communications technology, two elements are still in common use today—the telegraph key and the QSL cards on the wall.

thousands of nonhams acquired formal radio training to aid the war effort. The years following the war saw a mushrooming in the amateur population and the appearance of a wide variety of commercial equipment designed specifically for amateur use. As new technology was adopted, ham interests became more sophisticated and began to develop into diversified specialty areas.

For the ordinary person, the exotic appeal of Macao, a peninsula near China, would probably be sufficient reason to travel there. Not so with Dave Bell, W6AQ. For him, Macao was a choice DX location from which to operate an amateur station. That is why he and his forbearing wife found themselves en route to that Far East travel mecca laden with portable radio equipment. "DX" in amateur radio circles means distance, or contacts with distant spots—the more remote and isolated the better. DXers collect postcard confirmations of such contacts, just as stamp or coin collectors collect their specialty items. If there are few or no hams in a locale like Macao, then the thing to do, if you're a DXer, is to go there yourself and set up operations. If you do it during a popular ham contest when thousands of avid contesters are on the air, and especially eager for a rare contact such as Macao, so much the better.

After two days of contesting on the air using a friend's station, Dave found what it is like to be the only operator in a rare country. The only respite was during the brief moments when he flicked the switch so he could quickly consume a sandwich and a cup of coffee. The rest of the time was pandemonium.

Skylab 3 astronaut Owen Garriott, W5LFL, says: "My fascination with space began when I took up ham radio while I was still in junior high school. That led to a decision to pursue a career in electrical engineering and to do graduate work in propagation at Stanford University in California. There I helped monitor the first signals from Sputnik 1, the first spacecraft."

"To be an amateur in Pakistan," explains Parvez Bukhari, AP2PA, accounting for the fact that there are only 56 licensed amateurs in the entire country, "means one must be absolutely committed in order to vanquish the many pitfalls to be encountered. Until recently, for example, it required three years to receive a license."

U.S. amateurs are much more fortunate. Our government has long been a strong supporter of the Amateur Radio Service and has given it encouragement in many tangible ways. One excellent example of that encouragement is the Novice license, designed specifically to make it easy for anyone with any interest at all to become an amateur and get on the air. People as young as 5 and as old as 80 have become Novices. The examination is conducted by mail and administered by any amateur at least 18 years of age, not related to the applicant and holding a General or higher class of license. It consists of two parts—a code test of the ability to receive at 5 words a minute and a written, multiple-choice test with 20 questions covering very elementary radio theory, operating practices, and the regulations.

The newcomer to amateur radio has a wide range of choices in setting up a station. The cost can be $200 or $1200; and the space required can be just a table top or even a couple of shelves in a closet. An amateur radio station consists basically of three things: a transmitter, a receiver, and an antenna. With modern solid-state technology, the transmitter and receiver can be combined in one "pack-

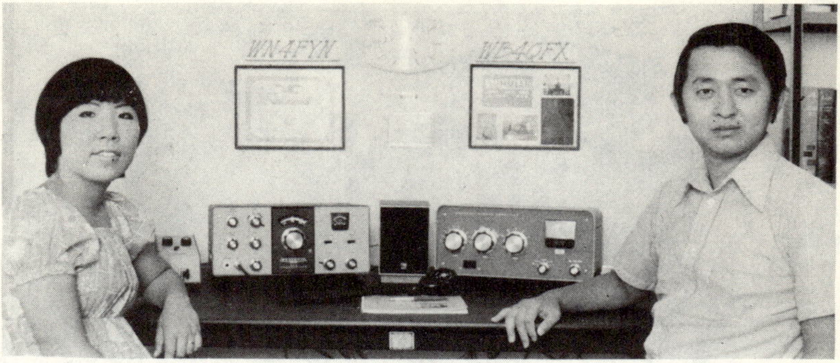

FIG. 2-3 One of the many couples who have joined the ranks of amateur radio operators. Their station is simple and uncluttered.

age" no larger than a shoe box—a transceiver. The most popular Novice antenna can be made from a couple of pieces of wire strung between two trees. Apartment dwellers have managed to get on the air with a piece of wire dropped out a window! Setting up a station is covered in greater detail in Chapter 6.

The first nongovernmental, nonmilitary space satellite went into orbit in 1961 as a tiny passenger on an Air Force rocket. It was designed and built by a group of California radio amateurs who christened it OSCAR, for Orbiting Satellite Carrying Amateur Radio. OSCAR 1 was essentially a scientific experiment which for a few days emitted a faint two-letter Morse code signal for "Hi." Today two OSCARs—much more sophisticated—are in orbit, both built by international teams of hams in several countries and both launched as "hitchhikers" with other satellites by NASA. They are space radio relay stations used by thousands of hams in more than 100 countries for casual conversation, for emergency messages, and for sophisticated experiments—and in a special classroom program bringing live space communication demonstrations to students everywhere.

"Let the skeptics say what they want about the code being an anachronism. I personally feel that it is actually an art form. It's a great feeling to get back to grass roots and listen to a fellow's personality, thinking and accent come through his key." That is how an old-timer explains the fascination of Morse code, the ham's special language.

Noted television entertainer and author Jean Shepherd,

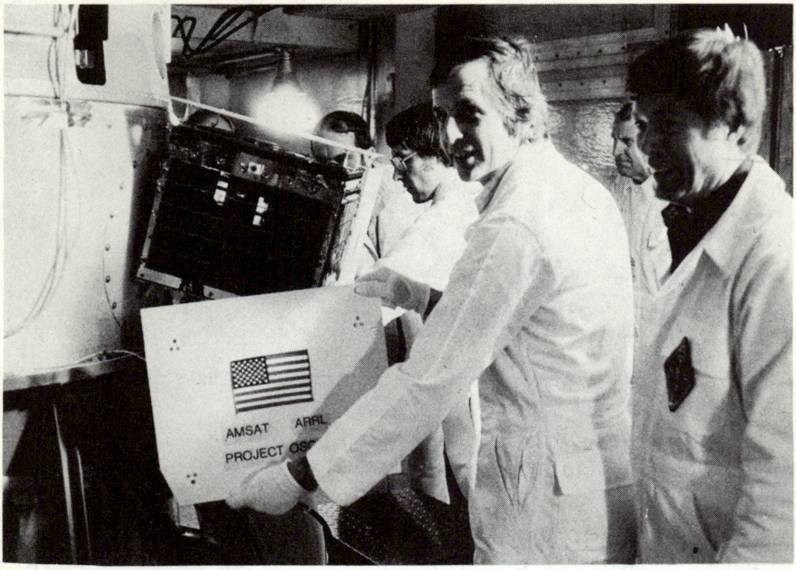

FIG. 2-4 Firmly attached to its launch vehicle, an OSCAR satellite gets a final checkout just prior to launch. Designed and built by amateurs from several nations, these satellites provide a fascinating means of entering the space age—from your home amateur radio station!

K2ORS, first became interested in amateur radio before he was in his teens. While he was still in high school, his ham license led to a part-time job at the local radio station, the first step in his broadcasting career. He is just one of the thousands of today's amateurs who credit their hobby as the impetus for careers in electronics and related fields. Arthur Collins, WØCXX, went from a youthful interest in ham radio to found Collins Radio Co., long-time supplier of high-quality military and amateur radio equipment. Altogether, about one third of licensed U.S. amateurs are electronics professionals or technicians.

Many hams are not scientifically or technically inclined, however. A young Connecticut woman, for example, got her first "ticket" simply because her husband was an ardent ham. In less than a year she had advanced to the highest-class amateur license, held by only 6 percent of the U.S. amateur population. The amateur ranks include housewives, teachers, students, lawyers, doctors, musicians, truck-drivers—and even a youngster who passed his Novice at age 5! W6SXR is Larry Wilson, warden at San Quentin prison. K6VFE is Sister Mary Charlotte of Fresno's Holy Cross Convent. W6NAZ is Lenore Jensen, popular radio and television actress. HK6LT is Luz Zuluaga, a former Miss Universe from Colombia, South America. WA1NHL is Betty Clay, Connecticut mathematics teacher. W1EFW is Milton E. Chaffee, retired bank president. VR6TC is Tom Christian, a direct descendant of the *Bounty*'s Fletcher Christian. W3ACE is Armin Meyer, former U.S. Ambassador to Lebanon, Iran, and Japan. AC3PT is former King Namgyal of Sikkim. W4RA is college professor Larry E. Price. K5YM is Texas real estate broker Tom Chance. W9PRN is Edmond Metzger, a theater chain executive. W4UG is Gay Milius, Virginia attorney.

Worldwide Friends, Despite a Disability

Otto Jarman broke his spine 16 years ago when he jumped into a reservoir to rescue a drowning child. Today, even though he is a paraplegic confined to bed, he is far from isolated. As an active ham he has friends in Mozambique, Puerto Rico, Nicaragua, and Senegal. Operating daily by the hour, he supplies travel directions, reports highway emergencies to the authorities, and chats with his friends. Some, like himself, are also disabled. He tells of one man who is unable to speak and can move only his big toe. "He sends code with that big toe and a straight key." There are also a former Montana highway patrolman, immoblized by a shot from a traffic violator; a Texan injured in a way almost identical to Jarman's accident; and many blind people. As Jarman says: "Amateur radio can take a disabled person out of his living room, out of his bed, out of his wheelchair and put him into the world."

High school UNESCO (United Nations Educational, Scientific and Cultural Organization) clubs around the globe (200 in Japan, 500 in France, and more than 100 in the United States) attract youngsters who want to "put the world more together" through international friendship. One group cruises the Pacific coast of California on an old wooden sailing sloop, the *Agua Allegre* (Merry

Waters). Aboard, young people swap ideas and cut through cultural boundaries. Using an amateur radio station, they have the entire world at their fingertips; conversations range from favorite snack foods and dating problems to pollution of the seas.

Andrew Osterberg, K1EPL, of West Yarmouth, Massachusetts, regularly "works" former neighbor Carl Chaplin, W7QO, now of Carson City, Nevada—but not with the usual voice or Morse code. They use slow-scan television. One evening Andrew had a mutual friend in his "shack" or radio room.

"Carl," he said, "do you know who this is?"

"Why, I sure do," came the excited reply from 2500 miles away. "Hello there, Bob."

Then Andrew said to anyone listening, "Anyone else get that?" Instantly from Hawaii came a voice, "A perfect picture." That was not particularly unusual; Andrew once "worked" a West German on slow-scan, a form of amateur operating that brings live pictures from around the globe into specially equipped ham shacks.

The closest ham counterpart to CB is called 2-meter FM. As a relatively short range means of local communication, primarily mobile like CB, it is almost ideal. There is far less congestion than on CB, yet certain frequencies are continuously monitored for emergency assistance, travel information, and the like. The range for 2-meter FM is considerably extended by devices known as repeaters, automatic relay stations that receive and retransmit radio signals. Usually built and operated by radio clubs or other local ham groups, repeaters are located on top of mountains, hills, tall buildings, or other high places (because a 2-meter FM radio signal is largely "line of sight," like a TV or broadcast FM signal). There are several thousand repeaters throughout the United States and Canada, and nearly all are listed in a directory published by the American Radio Relay League. Wherever they go, hams with 2-meter FM equipment in their cars always have access to someone. Nearly half of U.S. hams today are equipped for 2-meter FM. A complete 2-meter FM transceiver can be nearly as small as a pack of cigarettes. With a special adapter hams can even "patch" into the local telephone system to call home or contact the police during an emergency.

Suppose that you are in a friend's car, which has a transceiver. How do you tell whether the unit is for CB or amateur 2-meter FM?

FIG. 2-5 When disaster strikes unexpectedly, as it did in Wichita Falls, Texas, amateurs with 2-meter FM equipment in their cars volunteer to pass messages out of the area. But most often, amateurs use the compact gear for casual conversations, especially while traveling.

The easiest way is to listen to how users identify themselves. If they use a "handle" like "Rubber Duck" or "Red Falcon," or if they use a call sign like KGA1234, it is CB. If, on the other hand, you hear names instead of "handles," and calls like WB1CUJ or K1XX, you can be sure it is ham radio.

"KA6PVA This is WD3CYB"

Ham call letters have a special, almost mystic, significance to their holders. Partly that is because they are symbols of accomplishment, of membership in a select fraternity. But perhaps even more important, a ham call is unique. There are many Bill Smiths and there may be more than one Roderick Dunn, but there is only one WB1CUJ anywhere in the world. Ham calls are often much better known among hams than their owners' names. On meeting another amateur at a ham gathering, a ham may ask, "Have I ever worked you? What is your call?"

Ham call letters tell several things: the initial letter indicates the country (W, K, N, or A for the United States). The numeral identifies the section of the country—in the United States, for example, 1 is New England, 6 is California, 2 is New York or New Jersey, and so on. The two- or three-letter suffix is the individual identification. A short call sign—single-letter prefix, number, two-letter suffix in the United States—identifies either an old-timer or the holder of the highest class of amateur license. Relative newcomers to the ham ranks in this country can be recognized by a two-letter prefix

FIG. 2-6 This California ham proudly displays his amateur call sign on three different vehicles. Call-sign license plates alert government officials at the scene of an accident or disaster that the driver can provide vital and reliable two-way radio communications.

and a two- or three-letter suffix—such as KB1XYZ. There are some things call letters do *not* reveal: age, sex, race, national origin, and social or economic background, for example. No system is more truly democratic.

Ham radio often is a fine adjunct to other interests and hobbies. On the air, 'round-the-world chess games are popular with many hams. Evelyn Fox, WB9QZA, a 78-year-old YL (young lady—any female ham), regularly plays contract bridge with her American Association of Retired People group on 40 meters, a popular ham band. Darryl Dippel, WA5AAO, and three ham friends all like photography and science as well as amateur radio. On an expedition to Desert Bolson near Big Bend National Park in Texas, they photographed the Marfa light, a local atmospheric phenomenon of considerable interest to scientists. They used amateur radio to keep track of each other while hiking to separate stations. It was an excellent means of communication in the barren desert, otherwise a potential danger. Two young mountain-climbing backpackers, Richard Simpson, W6JTH, and Richard Dickson, WA6BVA, carried compact, portable, self-powered amateur equipment to the top of Mount Langley, where they made contact with hundreds of other amateurs, some as far as a thousand miles away. Spelunkers (cave explorers), model-railroad enthusiasts, sports-car rallyists, hunters, fishermen, campers—these and more use amateur radio to extend their hobby activities.

The Duke Medical Center Amateur Radio Club in Durham, North Carolina, through its club station, WB4BLK, performs a unique and often dramatic international public service. It operates a daily shortwave medical emergency service designated Project MED-AID. Ham radio operators in a number of Central American, South American, and other countries monitor their stations regularly; many remote mission stations have their own amateur radio facilities, and several mission doctors are amateurs themselves. This system brings an average of several calls a week for instant consultation with Medical Center experts about diagnosis, treatment, and prevention from field outposts and jungle hospitals. A doctor in Bolivia calls for information on tests for rapid checking of water purity to stem an outbreak of disease. A doctor in San Salvador not only gets help in determining treatment for a 14-year-old boy critically ill with kidney disease, but fast assistance in locating a special drug required and getting it on its way to him. A missionary in the Congo maintains direct contact throughout a night-long vigil with a North Carolina neurosurgeon to avert a crisis in treating a 2-year-old's serious head injury. A hospital director–general practitioner in Nicaragua has become a regular user of the MED-AID Service. He has consulted with Duke specialists about possible cancerous growths and has performed complicated surgery under their guidance. A doctor on the scene who suspected that the 11-year-old daughter of missionaries in Ecuador had leukemia contacted MED-AID; within 30 hours the girl had been flown to Duke. Through continuing improvement and expansion of the system, Project MED-AID is becoming an increasingly useful lifesaving service for isolated spots around the globe.

Sister Charlotte, K6VFE, uses amateur radio as an exciting adjunct to regular classwork. Her students at St. Agnes High School in Los Angeles were the first to learn of a new discovery during contact with an oceanographer working at the equator. He told them of his recent discovery of the *Neo paulina*, a species of fish previously thought to be extinct.

One obvious problem in considering any kind of international people-to-people exchange is the language barrier. The Peace Corps and multinational business firms solve that problem with accelerated langauge training programs. International travelers can work with language records or tapes. For hams it is much simpler. They have their own universally understood international "shorthand" system called *Q-signals*. "QTH," for example, means station location or home address—whether it be in Gambia or in Wyoming; "QSO" is an on-the-air contact; "QRM" is noise or interference; "QSL" is an acknowledgment of an on-the-air contact. A Q-signal can be either a question or a statement; "QTH?" means "What is your location?" There are other traditional ham symbols—such as "73" for "best regards"; "88" for "love and kisses"; "OM" ("old man"), a standard address for a male ham; "YL" ("young lady"), a female ham, who becomes an "XYL" after marriage. (The women's movement has brought some dissatisfaction with "YL" and "XYL.") "CQ" is an opening call on the air that means "I want to talk with any station." "FB" ("fine business") means good or excellent. A "rig" is station equipment. "XMTR" means transmitter; "XTAL" means crystal. With these signals and symbols hams may not be able to carry on an intellectual discussion, but they can and do establish basic communications.

Every day, hams by the hundreds in every corner of the world are in direct, personal contact with each other. It is an international, people-to-people exchange absolutely without parallel. Hams communicate by means of voice, Morse code, radioteletype, facsimile, and even television. A Russian ham on the air is not concerned with international politics but with the height of his American counterpart's antenna. From Kyoto may come an inquiry about how the Yankees are doing in the pennant race. A teacher in Kalamazoo will discuss classroom disciplinary problems with her counterpart in Calcutta. A Chilean attorney will report his experience with a new transistor circuit to the owner of a laundry in the Bronx.

Dr. Oswald G. Villard, Jr., W6QYT, pioneered studies of radio-wave propagation. Today, as director of the Ionospheric Dynamics Laboratory at Stanford Research Institute in California and a member of the National Academy of Sciences and the Space Science Board, he sees amateur radio as an underrecognized educational and self-development tool. "Ham radio, to me at least, has always prepresented a marvelous extra-curricular opportunity to see, feel and experience a fascinating form of electronic communication, all the way from initial composition of a message through acknowledgment of its receipt from a station on the other side of the earth."

Will You Be a DXer or a Contester?

Today's hams have a wide range of special interests. Some are builders and experimenters for whom getting on the air is incidental. Some are collectors and restorers of antique equipment who may look down their noses at modern gleaming cabinetry and shiny knobs. Some are contesters to whom sitting glued in front of a transceiver around the clock is just part of the game. Some like to *handle traffic*, which means relay messages as part of an extensive, cross-country on-the-air organization. Just about all are "rag chewers"; that is, at one time or another, they like to get on the air just to chat with another ham, any other ham. When a ham sends out his CQ, a general invitation to a contact, he never knows who may answer—or from where. That's part of the perennial excitement. A first on-the-air contact is generally acknowledged by an exchange of QSL cards, postcard-size written confirmations. QSL cards are highly personal and often highly individualized. In the course of a ham lifetime, an operator may collect hundreds or even thousands of these mementos. The more exotic will often be displayed on the walls of the "ham shack" or radio room. One of the most popular awards issued by ARRL, and certainly the most famous, is its DXCC certificate, awarded for verified contact with hams in 100 or more different countries. A "country" in ham terms may be England or France or Japan—or it may be a remote, uninhabited island in the middle of an ocean. It is all according to ARRL's Official Countries List.

Field Day is perhaps the single most popular ham event of the year. It takes place annually on a June weekend and in essence involves groups of hams setting up stations in the field, using portable power. The idea is to collectively accumulate the maximum possible number of contacts during the 48-hour time period. But that description does not begin to tell the story of Field Day. Conceived many years ago as an emergency-communications-preparedness exercise, it is still that today. It is also a combination picnic, outing, camping trip, contest, group activity, and chance to do a lot of operating under rugged conditions. It usually works like this. A local radio club sets up for Field Day at a campsite, on a hilltop, or in a town park where the public can watch amateur radio in action. The setup usually consists of several amateur stations and antennas for various bands and operating modes. A report from one participant is illustrative. "On Saturday morning of Field Day the group put up the antennas and hooked up the gear. The other station with more experienced operators would come by and check on us. Our group placed in the top ten in our category. But mainly, we had a good time. All the usual things happened. The soft drink supply ran low. Where do you buy soft drinks at three in the morning in a town with a population of under a thousand?"

In the United States, the FCC issues five classes of amateur licenses. An amateur license is really two licenses—an operator's license and a station license combined in one "ticket."

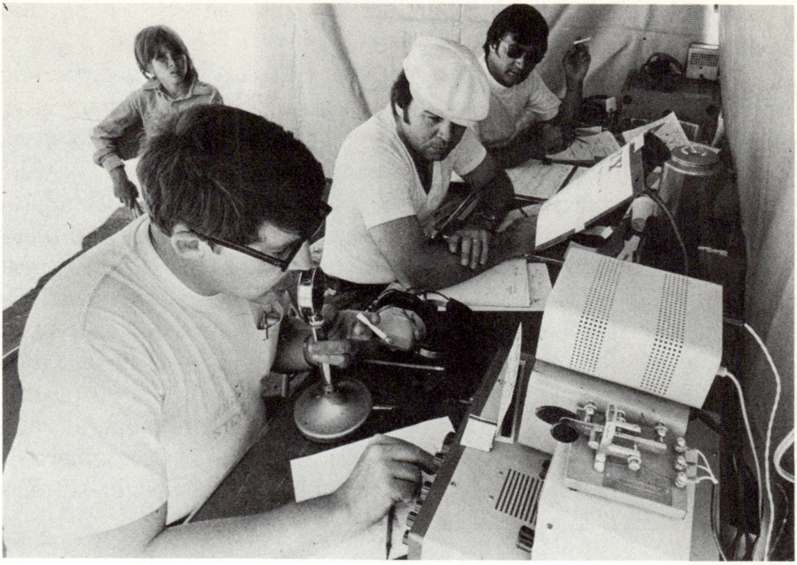

FIG. 2-7 Many thousands of amateurs "head for the hills" during the annual Field Day weekend in June. The idea is to prepare for an emergency by operating in the field, away from commercial power lines.

The first of the five classes is the *Novice*, created in the early 1950s to make it easy to become an amateur, to get on the air. It offers limited privileges: Novices may use only Morse code, and a maximum power of 250 watts. But all that is required is a 5-words-per-minute Morse code test and a short quiz that covers elementary knowledge of radio theory, operating practices, and regulations. Although it is possible to study on your own, many beginners choose to take a Novice class. To learn where one is being held in your area, contact the American Radio Relay League, Newington, CT 06111.

The *Technician* license is the next step up. (Incidentally, it is not necessary to progress through each license one at a time; beginners can start by taking the examination for the highest-class license if they wish.) The Technician exam includes the same code-speed requirement as the Novice—5 words per minute—and a more comprehensive written examination on radio theory, operating, and regulations. Technicians have more operating privileges than Novices do. In addition to Novice privileges, they may use voice in the VHF frequencies, which include 2-meter FM and the OSCAR satellite frequencies.

The *General* is the most popular class of license. It permits operation on all the popular ham bands using voice, code, or any other mode allowed amateurs. It requires a code proficiency of 13 words per minute and the same written exam as the Technician.

The *Advanced* class license has the same 13-word-per-minute code proficiency requirement as the General and a more difficult written examination. It allows operation on additional segments of all the popular ham bands, so Advanced class licensees have all the privileges of the lower classes plus some special ones of their own.

The *Amateur Extra* is the highest class of amateur license—the top of the line. It requires a 20-words-per-minute code proficiency and a highly technical written examination. It permits operation on all the frequencies allotted to amateurs, including some exclusive to Extra class operators. It entitles the holder to a two-letter-suffix or other preferred call sign if he or she so chooses.

Anyone Can Be a Ham

There are no age restrictions in amateur radio. About 10 percent of U.S. licensed amateurs are under 20 years of age; about 25 percent each are in the 20- to 35-year, 36- to 50-year, and 51- to

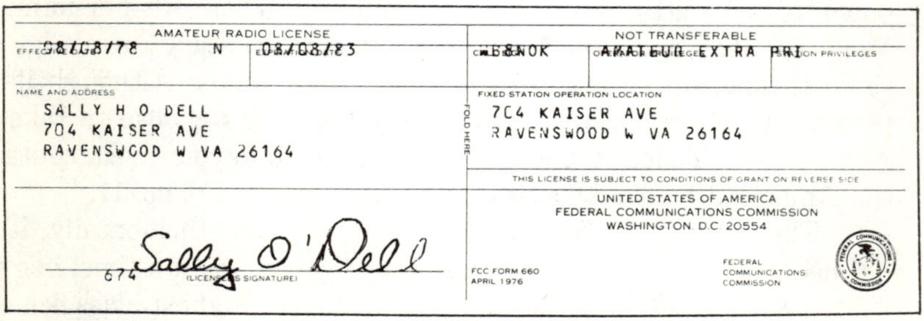
AMATEUR RADIO LICENSE
NOT TRANSFERABLE
EFFECTIVE DATE 08/08/78 N
EXPIRATION DATE 08/08/83
CALL SIGN
OPERATOR PRIVILEGES AMATEUR EXTRA
STATION PRIVILEGES PRI
NAME AND ADDRESS
SALLY H O DELL
704 KAISER AVE
RAVENSWOOD W VA 26164
FOLD HERE
FIXED STATION OPERATION LOCATION
704 KAISER AVE
RAVENSWOOD W VA 26164
THIS LICENSE IS SUBJECT TO CONDITIONS OF GRANT ON REVERSE SIDE
674 Sally O'Dell
(LICENSEE'S SIGNATURE)
UNITED STATES OF AMERICA
FEDERAL COMMUNICATIONS COMMISSION
WASHINGTON D.C. 20554
FCC FORM 660
APRIL 1976
FEDERAL COMMUNICATIONS COMMISSION

FIG. 2-8 The amateur radio license is a ticket to worldwide communications and friendships. In the United States, the Federal Communications Commission issues licenses to those who have passed tests in Morse code proficiency, radio theory, and rules and regulations.

65-year brackets; the remainder are in the over-65-year group. Among the 25,000 or so new amateurs each year, however, more than half are teenagers—or younger! About 55 percent of today's hams have attended college. About 40 percent are employed in electronics fields, one-third of those as engineers or technical employees and another third in professional or managerial capacities. Just about half of all U.S. hams are or have been citizens' band operators, and 4 out of 10 presently hold CB licenses.

How does one go about becoming a ham? How difficult is it? Is it expensive to get started?

Let's start with the last question first. Amateur radio can be compared to any other hobby: You can get started very simply or very expensively. A typical first station might cost about the same as a modest stereo system or a good CB rig. With brand-new equipment, you can get set up for about $350; with used gear, for under $150.

A person with no previous electronics or technical background can prepare for the Novice quiz with about 6 to 8 weeks of study for, say 4 hours per week. In many communities, local radio clubs or adult education programs offer low-cost licensing classes.

Excellent text materials are available from a number of sources. The American Radio Relay League offers a complete Novice package, containing a highly readable manual and a 1-hour, start-from-scratch Morse code tape—everything needed to obtain a Novice license.

For more detailed information on how to get started, write the American Radio Relay League, 225 Main Street, Newington, Connecticut 06111.

Is it easy to become a ham? Is it really worth it? Just ask any ham.

3

Amateur Radio: Some Basics

What *is* radio, anyway? How does "wireless" work? How can a voice or a sound slice through the air over many thousands of miles?

The answers to these questions are not only basic to what amateur radio is all about, but they open the door to some of the mysteries of broadcast radio and television as well.

An understanding of radio basics may be interesting to the broadcast listener or viewer; but it is *essential* to the radio amateur. Hams do not just listen (or receive); they transmit as well. To do that legally they must be licensed, and to do it effectively they must know what they are doing. Otherwise, the very busy radio spectrum would be a hopeless clutter indeed.

An amateur radio station has three basic and essential components: a transmitter, a receiver, and an antenna. Everything else is a refinement. The basic components can be very simple or very complex—representing an investment of as little as $100 or as much as many thousands of dollars. In that respect amateur radio equipment is much like photographic equipment. An inexpensive cartridge camera will take fine pictures under limited, ideal conditions which are entirely adequate for many people. To take pictures under all kinds of conditions and to provide creative flexibility, however, more sophisticated equipment is necessary. Just as avid photographers want faster, interchangeable lenses, high-speed film, filters, and an electronic flash, so dedicated hams seek out more powerful transmitters and more sensitive receivers, bigger and more elaborate antennas, plus auxiliary gear to enable them to work with greater ease and convenience. Yet, like photography, no matter how complex or costly the equipment, it all performs the same basic functions.

What a transmitter does, in essence, is generate radio-fre-

FIG. 3-1 Youngsters listening in on a ham conversation with simple-to-operate, inexpensive amateur equipment. Its few basic elements are easy to assemble and require only a modest investment, yet they serve the needs of beginning hams.

quency energy and modify that energy into an intelligible signal form. The antenna radiates the signal in electromagnetic waves much like the spreading ripples from a stone dropped into a pond—except that radio waves may emanate in all directions or they can be made to be directional, depending on the antenna design. Traveling at the same speed as light, 186,000 miles per second, radio waves may carry a signal in the form of voice, Morse code or video. That signal is picked up by a receiving antenna (often the same antenna used for transmitting) and passed into a receiver, which converts it back into an intelligible form and delivers it via a speaker or earphones as voice or code.

A radio signal gets from a transmitting antenna to a receiving antenna across town or around the world by a process called *propagation.* Incidentally, much of what is known today about propagation was first discovered by radio amateurs.

Depending on the means of propagation, radio waves can be classified as ionospheric, tropospheric, or ground waves. The *ionospheric* or *sky wave*, is that portion of the total radiation that leaves the antenna at angles somewhat above the horizontal. Except for the reflecting qualities of the ionosphere, it would be lost in space. The *tropospheric wave* is that part of the radio signal that is kept close to the earth's surface as the result of bending in the lower atmosphere. The *ground wave* is directly affected by the earth's surface. Depending on the frequency used and atmospheric conditions, radio waves can be reflected back to earth some distance away. This is called *skip.*

The portion of the radio spectrum that amateurs use most often extends from 535 kilohertz at the low end of the broadcast band, through 2000 kilohertz, or 2 megahertz, to 30 megahertz. This includes all the amateur "high-frequency" bands as well as bands used by other services. In the old days a signal was identified by its *wavelength* instead of its frequency, which is why hams still use the terms 40-meter band (7.0 to 7.3 MHz), 20-meter band (14.0 to 14.35 MHz), and so on. (The citizens' band channels 1 through 40 occupy the frequencies from 26.965 to 27.405 MHz.) The portion of the radio spectrum from 50 MHz to 3000 MHz is divided into groups known as the Very High Frequency (VHF) and the Ultra High Frequency (UHF) bands. Above those are the Super High Frequencies (SHF), also known as microwaves. (The commercial television VHF channels 2 through 13 occupy frequencies between 54 and 216 MHz; the TV UHF channels 14 through 83 occupy frequencies between 470 and 890 MHz.)

Various portions of the electromagnetic spectrum are assigned to specific services by national governments under international agreements. The International Telecommunication Union (ITU) is the responsible body. An agency of the United Nations, it is convened periodically at World Administrative Radio Conferences to review and reassign frequency allocations. WARCs usually take place about every 20 years. The Amateur Radio Service, through its own international organization, the International Amateur Radio Union (IARU), prepared for the WARC held in 1979 for several years. Amateurs have considered this effort particularly vital because of the increasing pressure for frequencies from other services and by the governments of the developing countries; many of the commercial, military, and other government services would like to expand their frequency allocations. Fortunately, many of the world's governments have a high regard for the value of the Amateur Radio Service in the national and public interest; the Amateur Service gained three new bands as a result of WARC '79.

Just after the turn of the century, the few hundred experimenters who called themselves amateurs were in the forefront of the rapidly developing frontiers of radio theory and technology. Hams were the first to discover the extended-range capabilities of the shortwaves, those below 200 meters, because they were relegated to these frequencies in 1912. At that time, the shortwaves were considered to be useless for all practical purposes, and the commercial and military interests of the day succeeded in having the amateurs confined to frequencies where they would not interfere with more important activities. But it was not long before they discovered that the shortwaves were in fact far superior for long-distance communication.

Most long-distance amateur activity is carried on by means of *sky waves.* The *ionosphere* acts as a great reflector of these waves because there, energy from the sun breaks up, or ionizes, the thin atmosphere into electrically charged particles. High-frequency radio signals cannot pass through the resulting layers and instead are reflected back toward the earth. The two principal ionospheric layers for radio purposes are at heights of about 60 miles and 150 miles above the earth's surface. How a radio signal is reflected off the ionosphere is determined by a number of factors: the time of year, the time of day, the frequency of the signal, and the angle at which it "hits" the ionosphere. The behavior of radio signals changes after sundown because ionization is caused by the sun. This explains why broadcast stations in distant cities can be heard after dark but not during daylight.

The smaller the angle at which a signal leaves the antenna and the higher its frequency, the farther away its reflected signal will be received. That returning signal may also be reflected back again from the earth and off the ionosphere. This process can, in fact, be repeated a number of times. With the right frequency and the right time of day, it is thus possible to reach very distant places. Radio waves tend to follow the *great circle* or most direct route around the earth, since they seek the shortest distance between two points. Thus, most radio communication is via the "short path," although long-path contacts do occur.

Many of the variables that affect long-distance radio communication are predictable, so every ham should have a basic understanding of them. From available propagation charts a ham can readily determine the *maximum usable frequency* (MUF) for a particular location and date.

It may be hard to imagine that spots on the sun can have anything to do with operating a ham station, but indeed they do. There is a sunspot cycle of about 11 years that affects ionization according to the intensity of sunspot activity. At the peak of the cycle, long-distance operating (DX) is enhanced. Frequencies otherwise usable only for short distances become effective for international contacts.

A Fascinating Gamble

DX is still a gamble, however—even for the ham with a sophisticated station, a big antenna, and a solid grasp of propagation theory. But that is a big part of its fascination. Sometimes when everything should be just right a signal will not get through; or it may

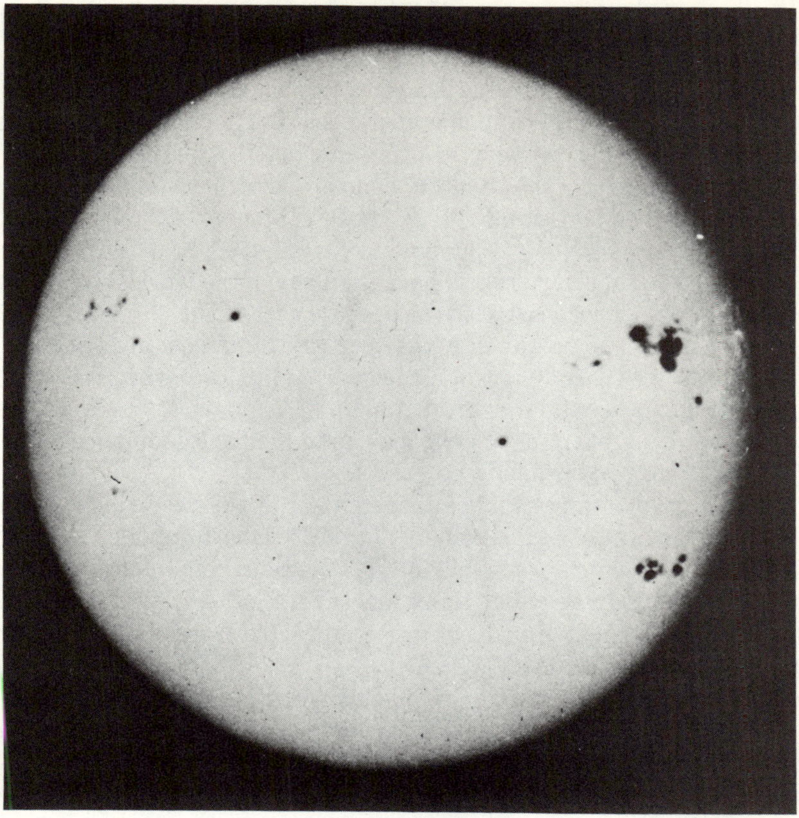

FIG. 3-2 Sunspots affect conditions in the earth's upper atmosphere, which in turn affect all electromagnetic communication. When many sunspots dot the surface of the sun, radio communication on earth is enhanced; at the low point of the approximately 11-year cycle, radio communication is degraded.

be received unexpectedly at a presumably inaccessible point. These situations can be caused by unusual eruptions on the sun, auroras in the polar region, or solar flares.

Novice licensees in the United States are permitted to operate, using Morse code only, on segments of four ham bands: 10, 15, 40, and 80 meters. Each of these bands has its own characteristics and its own advantages and disadvantages. Ten and 15 meters are the international bands, the DXers' favorites to reach the Far East, Africa, and Europe. But operating on these bands can be frustrating, with conditions often unpredictable. Some days when the band should be "open," it may be impossible to hear a signal; on other days, signals may come booming in from long distances. But for the new ham, 10 and 15 meters can be very exciting indeed, the places where the glamorous action is. The 40-meter band is one of the most useful U.S. ham bands. In the daytime it is great for contacts within the United States; in the evening it is possible to work some DX. This band can also have a lot of interference, because many of the world's commercial shortwave broadcast stations operate there. The 80-meter band is especially popular with Novices because it is almost always busy; after sunset there is always someone to communicate with. It is very reliable and is useful for short distances up to about 350 miles in the daytime. At night the range expands up to 1000 miles or more; in the winter 80 may occasionally be good for DX.

The radio signal that is transmitted to the receiving-station antenna can take one of a number of forms of radio-frequency energy. The two principle forms (modes) of amateur communication are Morse code and voice, but there are several others. These include slow-scan TV (SSTV), amateur fast-scan TV (ATV), radioteleprinter (RTTY), and facsimile (FAX).

Amateurs refer to Morse code transmission as *CW* or *continuous wave.* It is the simplest form of radio transmission, as well as the most efficient and most dependable. A CW signal occupies the smallest frequency space of any mode, requires the least power for reliable communications, and works best in noise. All of this explains why a Novice CW transmitter or receiver can be built very simply and inexpensively, and why CW proficiency is a licensing requirement of most of the nations of the world.

Code is still popular among hams. Newcomers to the amateur ranks enjoy it because it represents a new type of knowledge and skill and marks them as members of a special fraternity. Old-timers like to use code because of its simplicity and reliability, and because they enjoy the cultivation of a respected and admired skill. To an old-timer, users of CW have their own distinctive style. He or she can often identify friends unmistakably by their "fist," which can be thought of as a Morse code "accent."

Learning the code almost always proves to be far less an obstacle than many prospective hams expect. The proficiency required to pass the Novice test can be developed in 2 to 3 weeks of determined practice for an hour or less a day. Thereafter, skill grows naturally and unconsciously with use until the new hams find themselves "copying" code readily without even hearing the dits and dahs as such—they have begun to develop their own distinctive fists.

There is no question that voice, or phone as it is called, is the most popular and frequently used form of communication among today's hams. In the 1930s, hams relied on *AM (amplitude modulation)* phone for voice communication, essentially the same as AM broadcast radio today. After World War II, however, *single-sideband (SSB)* appeared on the amateur scene. It now shares voice communication largely with FM, which is used almost exclusively for local and mobile communication. SSB has taken over a large part of the on-the-air activity of modern amateur radio—from casual "rag chewing" or conversation, to contests and DX.

What *is* sideband? What does it mean?

A microphone transforms sound energy into corresponding electrical energy, but the frequency range that can be perceived as sound is only from 50 Hz to 20 kHz. To transmit this electrical energy, it must be raised to one of the amateur radio frequencies. This process is called *modulation*. In effect, it adds the voice frequencies to the transmitter frequency (or *carrier frequency*). But electronic theory normally requires that during modulation the audio frequency be both added and subtracted. If, for example, 3-kHz audio tone is modulated on a 3900-kHz carrier, it must also be both added and subtracted from 3900 kHz. The result is three frequency signals: at 3897, 3903 (the *sidebands*), and 3900, the carrier. Since both sidebands carry the message, one can be eliminated. This is exactly what happens in a single-sideband transmission.

The electronics required are obviously more sophisticated than for AM, and the equipment is therefore more complex (and expensive). But the development of SSB technology has changed the course of amateur radio. Since, as we have seen, SSB signals are narrower than AM, more signals can be transmitted in the crowded phone bands, and they are less vulnerable to interference. What's more, they are more efficient than AM, watt for watt, because the carrier and other sideband are eliminated. Although SSB is much

FIG. 3-3 Most hams prefer radiotelephone, or voice, communication. This mode of operation was greatly stimulated by the development of single-sideband after World War II.

more efficient than AM, it still uses far more frequency space than does a CW signal. A recent development, called narrow-band voice modulation, in effect cuts the frequency space for a voice signal in half and may have a significant effect on amateur radio in the future.

The other form of voice communication in amateur radio is *FM (frequency modulation)*. Hams use FM almost wholly for short-distance, mobile communication, primarily on the 2-meter band. This type of radio signal has become one of the most popular modes among hams. It is similar in application to CB, but since amateur radio has no distance restrictions, its range is considerably extended by relay devices called *repeaters*. Because they are located on mountaintops or other high places, repeaters receive and retransmit 2-meter FM signals over much greater distances than are possible with direct line-of-sight transmissions. Just about every community in the United States is within range of a repeater, so a traveling amateur always has access to at least one. A directory published by the American Radio Relay League lists most U.S. and Canadian repeaters.

Popular among a small, dedicated group of amateurs, *radioteletype (RTTY)* is a means of sending and receiving written messages over the air with a typewriter-like machine. Messages are printed out automatically at the receiving end in capital letters. A transmission can be received without an operator even being present, so a friend can leave an entire message for the recipient to read later. Messages are printed on a special tape which can be pasted on a message form like the old Western Union telegrams. In fact, many of the RTTY machines used by hams are Western Union discards, modified for amateur use. Many of today's RTTY enthusiasts use computers to send and receive amateur transmissions. These machines are fast and silent, in sharp contrast to the older type of RTTY device.

Two Types of Television

Amateurs can pursue television in two ways. *Amateur fast-scan television (ATV)* is essentially identical to commercial TV. Hams often buy used closed-circuit television cameras and modify surplus commercial UHF FM transceivers and standard black-and-white TV sets for this use. Amateurs use ATV on the UHF bands.

There is a problem with ATV, however; its signals take up more than 5 MHz of bandwidth, more space than in all the amateur bands below 6 meters. For anything other than purely local use, there has to be another way. That other way is *slow-scan TV (SSTV)*, which, as its name implies, is a TV signal with a very slow scan rate. A fast-scan TV signal produces 30 frames per second; in contrast, it takes 8 seconds to send just one SSTV frame. If normal TV is like a movie, SSTV is like a slide show. What makes SSTV attractive to hams despite its shortcomings is the fact that it can be used in any amateur phone band above 3.5 MHz. Anyone who can be contacted with a good

FIG. 3-4 Amateur slow-scan television (SSTV) produces a picture quite different from the familiar TV image. SSTV may be used on several amateur bands.

SSB signal can be worked via slow scan. At least one amateur has worked or made contact with more than 100 countries on SSTV. Operating SSTV requires only standard SSB equipment plus a monitor (the slow-scan TV set or receiver) and a camera. What is displayed on an SSTV screen looks like sections of the picture appearing slowly across the screen; the whole picture or image does not appear simultaneously. Focusing is a bit tricky and lighting is particularly critical in camera work because of the scanning and because of the limited brightness range. Since SSTV projects a series of still pictures, hams typically transmit images of themselves, graphic cards they prepare—sometimes very ingeniously—slides, and of course, shots of their station equipment. Newspapers and magazines are also a good source of material, as are family photographs.

Still another form of amateur communication that has a small body of specialized practitioners is *facsimile (FAX)*. It is similar to SSTV except that picture definition is far superior and it is registered on paper rather than on a picture tube. It can be used on any of the amateur VHF/UHF bands. The image is usually recorded on special light-sensitive paper wrapped around a drum.

Satellite communication is not, strictly speaking, a separate mode like SSB or SSTV, but it is used by hams as a specialized technique. And it has its own body of dedicated and enthusiastic followers. The first amateur satellite was designed and built by a group of California hams employed in the space industry. They named it *OSCAR*, for Orbiting Satellite Carrying Amateur Radio. It was launched as a "hitchhiker" with another satellite by the Air Force in 1961. The tiny, 10-pound box was essentially a scientific experiment—a demonstration of amateur skill and ingenuity. OSCAR 1 transmitted "HI" in Morse code for 3 weeks before its small battery was exhausted. Yet in that brief period of time, more than 5000 hams in 28 countries reported that they had heard the satellite. OSCAR 1 went into orbit just 4 years after Sputnik 1 brought the world into the space age. The Project OSCAR group in California, with the support of the American Radio Relay League and others, designed and developed three more OSCARs, which had increasing capabilities. OSCAR 5 was the first international amateur satellite, designed and built by students at the University of Melbourne under the auspices of the Wireless Institute of Australia. It was prepared for launch by the newly formed Radio Amateur Satellite Corporation (AMSAT), a group of hams in the Washington, D.C., area. AMSAT was formed in 1969 to carry on the work of the OSCAR group in California. OSCAR 5, launched in January 1970, transmitted signals on the 2- and 10-meter amateur bands. It was the first amateur satellite to be used for educational purposes. OSCAR 6, with a design lifetime of 1 year, supplied reliable operation well into its fifth year, despite an unprecedented level of usage by thousands of hams around the world. It continued to function effectively, in fact, far longer than the commercial satellite it was launched with in October 1972. Not only was it widely used for amateur communication, but it was the basis for what has become a highly successful science education program. In addition, it was used for experiments in disaster relief, medical data transmission, and intersatellite linking.

All the OSCARs since OSCAR 5 have been launched as auxiliary space "packages" by the National Aeronautics and Space Administration. NASA has done this in recognition of their scientific and educational value as well as their public service potential.

OSCAR 7, launched in November 1974, carries two solar-powered transponders or radio relay stations. Its power is 1 to 2 watts. OSCAR 7 is octahedral in shape to provide maximum exposure for its solar panels.

OSCAR 8, launched in March 1978, also has two transponders aboard. The first of a series of amateur satellites having highly elliptical orbits, AMSAT-Phase III will be within range for many hours at a time, far longer than the 20 minutes or so of previous OSCARs.

With orbital data supplied by most of the amateur radio periodicals, amateurs can predict the exact time and point at which OSCAR will come within range of their station location. Anyone can listen to OSCAR's signals on an amateur-band receiver that covers the 10-meter and 2-meter (SSB and CW) bands; a Technician class or higher license is required to transmit through the satellites. Most popular, general-coverage shortwave receivers include the 10-meter *downlink* or receiving frequencies. A simple wire antenna strung to a tree is all that is necessary to hear OSCAR clearly. With the appropriate equipment, hams use voice, Morse code, and even slow-scan television to transmit signals up to 10,000 miles via OSCAR.

A new dimension was added to amateur satellite communications when two Soviet amateur satellites, designated RS-1 and RS-2, were launched in the fall of 1978. Both were similar to OSCAR 6.

The OSCAR satellites allow just about any ham to become an instant participant in the space age with ordinary, inexpensive amateur equipment.

Perhaps the greatest source of initial confusion for the new or prospective ham is the equipment involved. No two ham stations are exactly alike, as amateurs have decided preferences when it comes to specific pieces of gear. The variety of transmitter, receiver, or transceiver types, antenna designs, and accessory equipment is seemingly endless. That variety is a reflection not only of differences in the size of ham pocketbooks, but to a much greater extent, of differences in their special interests. Amateur radio is very much in an era of specialization, and that is the real key to assessing equipment needs.

Receivers

There are two basic types of receivers used by hams: general-coverage and ham-bands-only. A general-coverage receiver spans a wide range of frequencies, usually between 0.5 and 30 MHz, so that domestic and international broadcasts, marine, aircraft, police, weather, and other services can be heard. The principal disadvantage of this type of receiver for amateur use is that it is relatively difficult to tune accurately and thus is not as convenient to use. There are usually two dials: a main dial for tuning first and a band-spread dial to zero in more precisely on a desired frequency. A ham-bands-only receiver covers just those frequencies amateurs use. For this reason it is easier to tune accurately and its dial is much easier to read. For equivalent quality, a ham-bands-only receiver is generally less expensive than a general-coverage unit.

Several kinds of tuning dials are found on amateur receivers. The easiest to read is the slide-rule dial. Not only does it show the frequency it is set to, but also where that frequency falls in the band covered. Its length, however, is limited by the dimensions of the receiver itself. A circular dial eliminates this problem, but only a

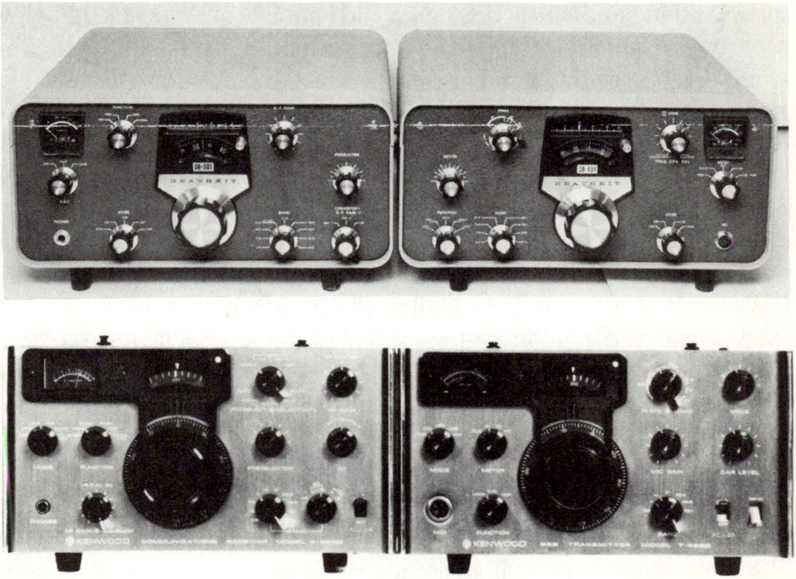

FIG. 3-5 Matched receiver/transmitter pairs are preferred by many amateurs for their operating flexibility. Some, like the Heathkit gear, are available on the used market for reasonable cost. Other, more sophisticated transmitter/receiver combinations, such as the Kenwood pair, boast smaller size, more features to facilitate operating, and sophisticated circuitry.

portion of the band is visible at one time. A popular dial mechanism combines the advantages of both the slide-rule and circular dials; it has a slide-rule dial for locating the desired frequency within the band and a circular dial for fine tuning. Digital frequency readout is a relatively recent innovation.

The most important specifications in an amateur receiver are for selectivity, sensitivity, and dynamic range. *Selectivity* is the ability of a receiver to separate two closely spaced signals, *sensitivity* is its ability to detect very weak signals, and *dynamic range* is its ability to perform in the presence of very strong signals. All are highly desirable, but for the new ham, selectivity is probably most important. The ham bands generally, and the Novice frequencies in particular, are congested, with many contacts occurring close together. This is especially confusing to one accustomed to the clear, uncluttered nature of broadcast radio. So the ability of a receiver to isolate a desired signal is a very attractive quality. Selectivity is measured in bandwidth; if a signal 3 kHz above and one 3 kHz below the tuned frequency can be heard, the selectivity is 6 kHz. Thus, the smaller the specified bandwidth, the more selective the unit. Selectivity is accomplished by filters built into the circuitry; for Morse code, a selectivity of about 400 Hz is considered good; for SSB, 2.3 kHz or so will do. Many receivers have a control that changes bandwidth.

Another feature every ham receiver should have is a BFO, or beat-frequency oscillator. Without a BFO there is no way to receive CW clearly. If there is no control on the front panel marked "BFO" or "CW," ask the person selling the unit or a knowledgeable ham whether the receiver has a BFO.

Transmitters

The transmitter is the heart of an amateur radio station; it is the difference between being a listener and a communicator. Transmitters come in all sizes, shapes, and costs. As every old-timer knows, how much hams spend for their transmitter depends more on what they want than on what they need. A basic consideration is whether the transmitter is to be used for CW only or for voice and other modes. CW involves the simplest electronics and is therefore the simplest and least expensive type of unit. There is a big jump in both cost and complexity from CW-only to single-sideband. Today's hams often want solid-state (transistorized) equipment. As a result, superb "old-fashioned" tube-type gear is sold relatively inexpensively. Such equipment tends to be heavy, bulky, and cumbersome by contemporary standards, but it can function very dependably. Solid-state gear, on the other hand, is compact, light, and portable. Because it does not generate the heat of tubes, it tends to be long-lasting and relatively maintenance-free. Where space is a consideration, an entire solid-state station can be assembled in a small fraction of the area required for tube-type equipment.

A crystal-controlled transmitter has a separate plug-in quartz crystal to select each operating frequency. This is often inconvenient and can involve maintaining a sizeable stock of crystals that may still not cover all operating needs. Most amateurs, therefore, purchase a transmitter with a built-in variable-frequency oscillator (VFO), which permits shifting to any frequency in a desired band.

Another consideration involves transmitter power. U.S. Novices are limited to a maximum of 250 watts of input power, but General and higher-class licensees can use as much as 1000 watts. Actually, many active and successful amateurs habitually use far lower power levels than the maximum allowed. There is even a growing body of devotees to *QRP*—operation at power levels below 10 watts. These specialists have achieved astonishing results, making contacts at great distances; the emphasis being, of course, on operating skill.

Transceivers

The ultimate in a compact amateur station is the transceiver, which combines a transmitter and receiver in the same "box." Two-meter FM and other types of gear designed for mobile use are ordinarily supplied only in transceiver form. A transceiver has certain advantages for all applications, in addition to compactness. For one thing, it is more convenient to operate just one set of controls for transmitting and receiving. The disadvantage is that the transmit

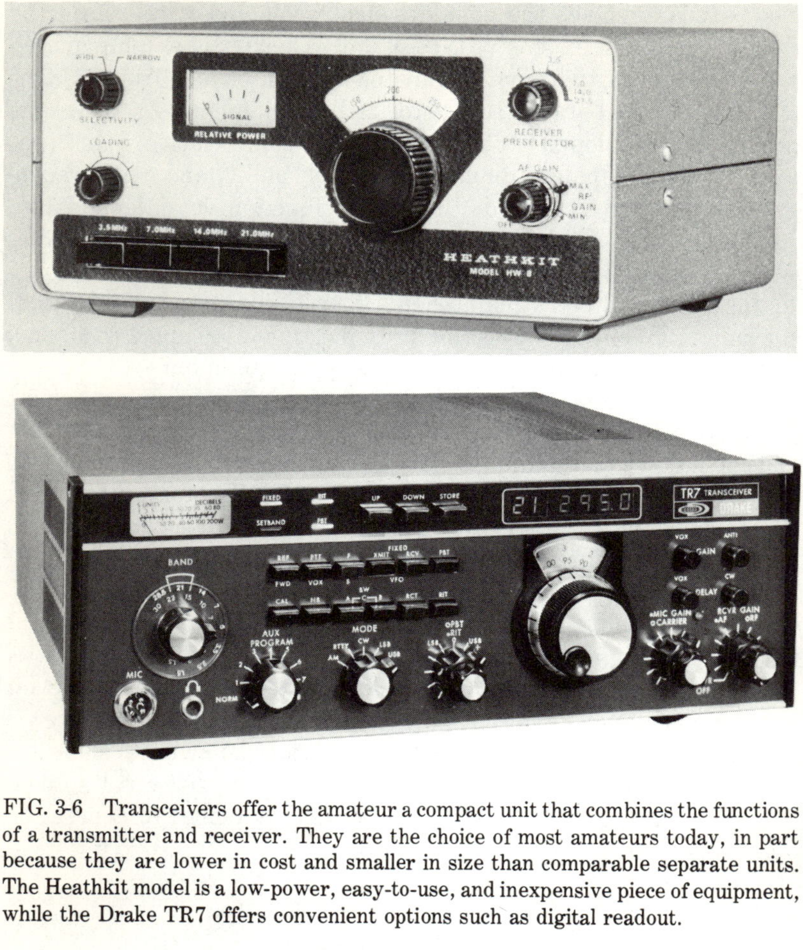

FIG. 3-6 Transceivers offer the amateur a compact unit that combines the functions of a transmitter and receiver. They are the choice of most amateurs today, in part because they are lower in cost and smaller in size than comparable separate units. The Heathkit model is a low-power, easy-to-use, and inexpensive piece of equipment, while the Drake TR7 offers convenient options such as digital readout.

and receive functions are locked together in frequency. If a significant difference in the transmit and receive frequency is required (known as *offest operation*), transceivers will not allow contact to be made. A transceiver generally costs somewhat more than either a transmitter or a receiver of equivalent quality, but less than both. The trend in amateur radio today, as a review of the advertising in amateur publications will demonstrate, is clearly in the direction of transceivers, but is far from a universal trend. Many hams would feel severely handicapped if they could not make use of separate transmitter–receiver combinations.

Antennas

Experienced hams consider the most critical element of an amateur station to be the antenna; it is the real key to successful operation. It is the antenna design, construction, and location that determine how effectively a signal is put out and whether it will come booming in over all others at a distant receiving point. Old-timers are fond of saying: "If your antenna did not blow down last winter, it is not high enough." Indeed, height is an important factor, and hams may try to select a home location on a hill or at least a relatively high, unobstructed place. One affluent ham purchased a new home sight-unseen. The seller was a well-known DX operator who was known to have a fantastic antenna array. What the purchaser bought, of course, was the "antenna"; the house was relatively incidental.

Ordinarily, the same antenna is used to transmit and to receive, although receiving can be accomplished much more simply than transmitting, even with a simple wire antenna. Antennas offer hams their best opportunity to be creative these days; they are the one piece of equipment that is most often homemade. Ham ingenuity has to take into account landlord restrictions and property size and location, as well as neighborhood relations.

For the beginning ham, three types of antennas have proven to be most popular: the half-wave dipole, and quarter-wave vertical, and the random-length wire.

The dipole is probably the most popular because it is the easiest to construct and will work just about anywhere with just about any kind of installation. It consists of two lengths of antenna wire, each equal to one quarter-wavelength at the intended operating frequency. The outer ends are suspended from trees, poles, or any other high support; the inner ends are connected at the center to a coaxial cable that brings the signal from the transmitter. The two wires do not have to be parallel with the ground; the center can be high and both ends lower (an inverted V), or one end can be high and the other low (a sloper), just as long as at least one portion is as high as possible.

FIG. 3-7 Whether they're arguing their merits, building them, or just reading about them, hams love their antennas. These antenna aficionados are in the process of raising a beam, the antenna preferred by most serious hams.

The length of the wire can be calculated by using the formula: length in feet equals 468 divided by the intended frequency in megahertz.

One problem with the dipole is its length. For the Novice 80-meter band, for example, the total length required for the dipole is about 130 feet. Additional bands can be accommodated without erecting separate dipoles. Pairs of wires for each band can be fed into the same central point or coaxial cable, or a device called a *Transmatch*, or antenna tuner, can be used to adjust the antenna characteristics electronically.

The principal advantage of the quarter-wavelength vertical antenna is that it requires little space. The antenna is merely a long, metal pole. At full size its length is equal to one-quarter the wavelength it is to be used for. For the 80-meter band, that means an antenna about 60 feet high. Fortunately, commercially manufactured vertical antennas are physically shorter than a full quarter-wavelength. Several manufacturers market antennas that are about 20 feet long and cover all four Novice bands, including 80 meters.

The key to successful operation with a vertical antenna is a good ground system. The best system uses many wires laid out like the spokes of a wheel from the base of the antenna. If placed aboveground, these "radial" wires must be about one quarter-wavelength in length; if they are buried (no more than 1 inch below the surface), the length is less critical. A multiband, aboveground installation requires separate radials for each band; there may be as many as 100 individual wires, although as few as four radials for each band may be sufficient. Vertical antennas can be mounted on a piece of pipe driven into the ground, placed on the chimney of a house, or mounted on a tower. Usually, the higher an antenna is mounted, the better it will work.

The third type of antenna popular with beginners is the random-length wire. It should be as long as practical, which for an apartment location in the city may be along the ceiling of a few rooms or in an attic; or it can be stretched out a window to a tree or pole. Since such an antenna may be of any length, it must be matched electronically rather than physically to the appropriate wavelengths by means of a Transmatch. This device can be adjusted for each of the different amateur high-frequency bands.

Sending Keys

The traditional key used for sending Morse code is the straight key, still preferred by some CW operators and used by most beginners to master code. It is basically an electrical switch that opens or closes the electrical circuit to form the code dits and dahs (short and long sounds). Spacing of the contacts and the tension of the spring that "opens" the key are both adjustable to individual operating preferences. There are two other kinds of keys for more advanced use. The "bug" operates mechanically but from side to side instead of vertically, forming automatic dits in one direction and dahs in the other. An *electronic keyer* is particularly useful for high-speed sending, up to 60 words a minute. Its circuitry automatically produces correctly spaced dits and dahs.

4

The Evolution of Amateur Radio

Early on an October morning in 1933 I was among a small group of people who gathered at the Dearborn Street station in Chicago to greet Marconi who was to be a guest at the "Century of Progress." A modern amateur radio station was one of the exhibits at the fair and was shown to the great inventor. The individual in charge, probably for lack of better words, made some apology for the equipment being built by only an amateur. It was then, I believe, that the "father of radio" made his often quoted statement: "Don't forget, young man, that I too am only an amateur" (Lewis Coe, W9CNY).

Marconi did not, in fact, "invent" radio. What he did was to develop the first practical application for a body of knowledge that had been growing for some years.

James Maxwell, an Englishman, predicted the existence of

radio waves in 1864, but it was not until 1888 that Heinrich Hertz, a German physicist, actually demonstrated that these waves existed and traveled through space. (Our debt to that eminent gentleman is marked today by our reference to frequency measurements as "hertz" rather than "cycles per second," as they used to be known.) Later, Ernest Rutherford, an English physicist, succeeded in sending radio signals three-fourths of a mile, and Oliver Lodge, another Englishman, developed the basic principles of tuning.

Marconi experimented with radio in Italy at the same time as Hertz in Germany, and they made many similar discoveries independently. Marconi left Italy and went to England, where radio research was expanding rapidly. He took out his first patent there in 1886.

Then, on a raw December day in 1901, Marconi made history when he sat in an old barracks at St. John's, Newfoundland, and

FIG. 4-1 Onlookers were skeptical when radio experimenter Guglielmo Marconi set up this imposing array of antennas. His goal was to send a crude type of radio signal across the wide Atlantic. Although this set of 200-foot-high antennas blew down in a nor'easter in 1901, Marconi persevered. (Photo courtesy National Park Service, Cape Cod National Seashore.)

launched the Age of Wireless. Across 2000 miles of space from Cornwall, England, he heard in his headphones a crackling series of buzzes—the letter "S" in international Morse code. That prearranged signal was a magnificent culmination of years of experimentation by many people, and it was instant worldwide headline news.

In 1922, when Marconi came to New York to deliver a lecture, he met a young radio amateur named Paul Godley, who held the call letters 2ZE, and invited him to dine with him aboard his yacht. The next day Godley assisted in the demonstration.

In the Smithsonian's Museum of History and Technology today, there is a 1915 vintage radio receiver designed by Godley. His famous Paragon was the first amateur receiver marketed as a complete unit; it revolutionized amateur radio because it adapted Major Edwin Armstrong's pioneering "regenerative" circuit to the amateur short waves. Godley's Paragon receiver was also used by the Army Signal Corps in World War I, the only American radio equipment to reach the Western Front.

For a number of years, of course, amateurs had themselves envisioned transatlantic shortwave transmissions, and by early 1921 a group of American and British amateurs was trying unsuccessfully to accomplish the feat. In November of that year, under the auspices of the American Radio Relay League (ARRL), the amateur organization co-founded in 1914 by another famous inventor, Hiram Percy Maxim, Paul Godley sailed for England to execute a plan proposed by the League's then traffic manager, Fred H. Schnell. Godley set up his station at Ardrossan, Scotland, and at 1:33 A.M. on December 8, 1921, recorded in his log the first reception of an amateur shortwave signal from the United States. Before the series of transatlantic tests had ended, more than 30 American amateur stations were heard by Godley and a group of British experimenters. Godley's log of those tests is still preserved in the museum at ARRL headquarters in Newington, Connecticut.

"The world knows and honors the Marconis, the Armstrongs and the Goddards," notes Arthur A. Collins, founder of the Collins Radio Company. "But radio amateurs like Paul Godley, John Reinartz, Mike Villard, Grote Reber and many others have discovered new things about radio and radio waves—not only earthbound, but in the ionosphere and in space as well. Amateurs have found new uses for communication science from the beginning."

FIG. 4-2 Noted inventor Hiram Percy Maxim, who held the call sign 1AW, founded the American Radio Relay League with another young man in 1914. Today, the League's 160,000 members represent the wide range of talents and interests of the 1 million hams worldwide.

John Fleming, an English electrical engineer, developed the first vacuum tube in 1904. It was improved by Lee De Forest, an American inventor, in 1906, opening the way to voice communications by wireless. The first radio broadcast was heard on Christmas Eve of 1906 when radio operators on ships at sea suddenly heard a man speaking, then a woman singing, and a violin playing, followed by, "If you have heard this program, write to R.A. Fessenden at Brant Rock."

The typical amateur station of those days was an induction coil, a condenser and spark gap for the transmitter, and a simple coherer–decoherer of galena crystal for the receiver/detector—usually feeding into a single head telephone. Better equipped stations had receiving tuners (most U.S. commercial receiving equipment was untuned, since patents on the loose-coupler system of tuning were held by Marconi). Although the Fleming valve had been

FIG. 4-3 The typical amateur station of 1915 sported equipment that was crude by today's standards. (Photo courtesy Antique Wireless Association Historical Museum.)

invented in 1904, and the De Forest audion in 1906, neither type of tube found immediate general acceptance in wireless communication, especially by amateurs, because of the cost. Distances ranged up to several hundred miles for the larger stations with several kilowatts, but for the most part, hams were content with 50 or 100 miles with average gear.

Chaos on the Airwaves

Regulation was nonexistent; there was no radio law. The Navy did issue "certificates of proficiency," but having one was not a requirement for operation. Everyone had an untrammeled right to the airwaves. Inevitably, conflicts developed between amateurs and military and commercial interests. Amateurs were accused of cluttering up the air and interfering with important traffic. Part of the trouble was that amateurs often had better and more powerful stations and were usually better operators. The military and commercial interests soon took their battle to the Congress, and their efforts eventually culminated in the Radio Act of 1912, signed into law by President Taft. At last, everyone felt, the "problem" of those amateurs was solved. Everyone believed they had effectively been put off the air because they were now restricted to the use of the wavelengths below 200 meters. It was the expert consensus of the day that those frequencies were worthless. Everyone knew that.

Everyone knew it except the amateurs. The law turned out to be one of those great blessings in disguise. Under the pressure of necessity, it did not take ingenious and determined amateurs long to discover that the "useless" shortwaves were in fact superior for long-range radio communication.

By the end of 1915, amateurs were once again accomplishing what were in those days unbelieveable feats in transmission and reception. With homemade equipment, often built for $100 or less, and in the despised 200-meter range, hams again outperformed government and commercial installations representing investments of thousands of dollars. It took hams only three years to again demonstrate superior technical and operating ability. Few achievements since in the field of radio electronics have really surpassed the amateurs' "discovery" of the short waves.

With the outbreak of war in Europe in 1914, hams quickly recognized the threat of engulfment to the United States. Through their fledgling amateur organization, the American Radio Relay League, and its cofounder-president, Hiram Percy Maxim, amateurs offered their services in the event of emergency situations. Charles Apgar, 2MN, using an ingenious adaptation to his receiver, gathered evidence that resulted in the government's shutting down the German radio station WSL on Long Island for neutrality violations. It was described as "the most valuable service ever rendered by a radio operator to this country."

In April 1917, by executive fiat from Washington, all amateur stations were directed to shut down their operations, both transmitting and receiving, for the duration of the war. When the United States went into the war the armed forces were faced with a total lack of qualified radio personnel. It was remarkably fortuitous that at the very time thousands of trained operators were so sorely needed, there were more than 6000 amateurs in this country who had been in training for years for exactly the sort of activity required.

Amateurs Respond to Call for Radio Operators

At the request of the Navy, the ARRL was able to recruit 500 operators in less than 10 days. A second call for 2000 operators was filled almost as quickly. Altogether it is estimated that nearly 4000 hams saw military service during the war. At least 50 were placed in positions of responsibility immediately and directly on the basis of their amateur experience. One was Lieutenant Clarence D. Tuska, cofounder of ARRL and its secretary, who received a direct commission and was placed in charge of organizing radio training in the Air Service of the Army Signal Corps. Captain (later Major) Edwin Armstrong, inventor of the Armstrong regenerative circuit used by every belligerent in the war, was put in charge of the Signal Corps Radio Laboratory in Paris. It was there, in fact, that he invented the superheterodyne receiver, a system so advanced for its time that it is still in virtually universal use.

U.S. amateurs supplied the nucleus for the most efficient wireless signal corps of any of the combatant nations. At the end of hostilities, the Secretary of Commerce wrote, "The officers in charge of the wireless operations of our armies in France commend highly the skill, ingenuity and versatility of the licensed amateur radio operators who volunteerd in large numbers for military service and served in dangerous and responsible positions."

Despite that record of wartime service, getting back on the air after the war ended proved to be a formidable task for amateurs. A bill to give the Navy control of all radio was defeated before the year of the Armistice had passed. In April 1919, the ban on receiving was lifted, but transmission was still prohibited. In July, another effort was made by the Navy to gain control, this time of international radio. This, too, was finally defeated. At last, under prodding from the ARRL, the Congress in a joint resolution directed the Navy to lift the transmitting restriction on amateurs, and the Navy complied in September 1919. Amid widespread rejoicing, hams were back on the air 2 years after operations were shut down.

At this time amateur radio was exclusively telegraphy, international Morse code. Although principles of telephony, including single sideband, the most spectacular post–World War II development for amateurs, were known and understood as early as 1919, hams used spark-gap transmitters that consisted essentially of a coil, an adjustable gap between two electrodes, an antenna, and a ground. A condenser was just an added refinement. The jumping of the spark between the two electrodes was always an impressive and sometimes disconcerting phenomenon. U.S. Sen. Barry Goldwater, K7UGA, has said: "It could really hit you on the head if you were not careful." Although 1 kilowatt might be good for transmitting a distance of 20 or 30 miles, most spark-coil operators were happy to be able to reach their buddies a couple of city blocks away. Just having a signal heard was tremendously satisfying. Distance could come later.

Come it did—and not much later. Within a couple of years communication over several hundred miles was commonplace. One of the major reasons for that rapid development was Armstrong's regenerative audion circuit, which vastly increased receiver sensitivity and selectivity. The effect was revolutionary, because spark-gap-transmitted signals were without amplification and covered a lot of spectrum space. If someone nearby was on the air, everyone else just quit unless they wanted to listen.

CW Replaces Spark

It took the development of power tubes to bring about the next big step forward in amateur radio technology: CW or continuous-wave transmission for telegraphy. The advantages of CW were considerable. For one thing, it caused far less interference to broadcast reception than spark did; an operator did not have to wait for the neighbors to go to bed! By the end of 1921 and early 1922, with the appearance of affordable and efficient commercial vacuum tubes, CW was on its way to acceptance by amateurs, and spark faded out long

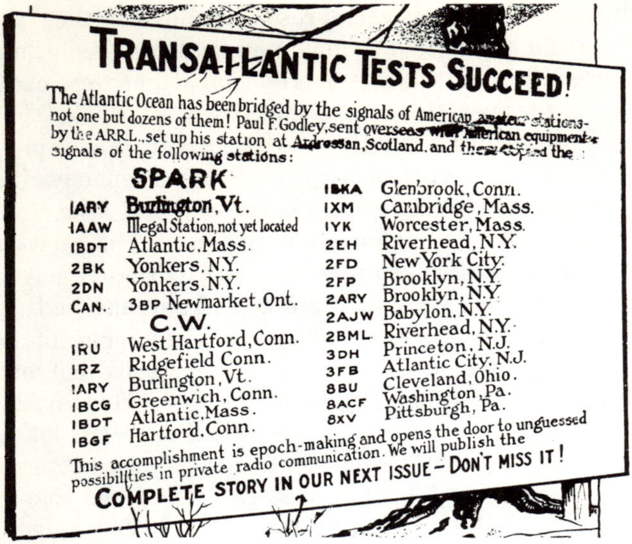

FIG. 4-4 The January 1922 issue of *QST*, the ARRL membership journal, displayed this list of North American amateurs who had communicated with Paul Godley in Ardrossan, Scotland.

before its legal demise in 1927. A big problem with CW, the difficulty of tuning, of keeping a signal audible, was solved with John Reinartz' invention of a receiving tuner. Reinartz, who held the amateur call 1QP, made many notable contributions to radio in the early days. He was one of those who first recognized that a reflecting layer in the upper atmosphere—the ionosphere—was responsible for "skip," the ability of shortwave radio signals to be received over great distances.

During the early 1920s, amateurs were urged to adopt "quiet hours" in the early evening to avoid interference with commercial broadcasts. It was one of their first forms of effective public relations. There was even a brief period in 1922 and 1923 when quiet hours were made compulsory by the Department of Commerce.

It is often thought that voice transmission or radiotelephony came along much later for amateurs. In fact, in 1922 and 1923, Ballantine's *Radiotelephony for Amateurs* was something of a ham bible. By 1932 "phone" was in reasonably common use by hams and frequencies were allocated for phone use.

By June 1920, 5719 amateur stations were listed by the U.S. government. By 1929 there were 16,000 amateurs, and by 1933 that figure had mushroomed to 41,555.

How broadcast radio evolved from amateur radio is interesting. In November 1920, Frank Conrad, 8XK in Pittsburgh, one of the first successful amateur CW stations, began the transmission of music instead of the dits and dahs of the Morse code. From that grew the present-day KDKA and a modern, billion-dollar industry. Actually, the *Detroit News* claims the first broadcasting station, WWJ, which started with a small De Forest "Oscillion Radiophone" 11 weeks before Conrad's station. But it was to hear 8XK that people first purchased radio receivers. Soon broadcast stations KYW at Chicago and WBZ at Springfield, Massachusetts, were established. WJZ came on the air in New York City late in 1921. Then, as now, amateurs were liberally represented on broadcast station staffs.

Amateur Radio Public Service

One of the proudest and longest traditions in amateur radio—and the basis for much of the governmental support it has received over the years—has been its spontaneous provision of communications services in time of emergency. The first record of such service occurred in 1913, when amateurs at the University of Michigan and Ohio State University provided the only available communications during a violent Midwest windstorm. By 1923 there were regular reports of amateur emergency communications services. In 1925, newspapers around the world headlined the entrapment of explorer Floyd Collins in Sand Cave, Kentucky. Communications between the rescue site and the nearest telegraph office were supplied by amateurs 9BRK and 9CHG, who operated continuously without sleep for 4 days because no one else was available to assist. In January 1926, the Pennsylvania Railroad requested that a special amateur circuit be set up to serve their system in emergencies. Many hams participated and for years the distinctive call "PRR" was a rallying call for amateurs serving the system. In Canada in November 1925, 4CG at Selkirk, Manitoba, spent 3 days trying to get medical aid from Winnipeg for a critically ill woman and child. Finally, through 9EBT in Fargo, North Dakota, contact was made and a doctor dispatched to the isolated village in time to save two lives. After a particularly vicious hurricane hit Florida in 1926, all wires and power lines were down and communications were nonexistent. Dozens of amateurs with calls like 4KJ, 4HZ, 4PU, 4SB, and 4IZ bridged the gap with their own battery-powered equipment. In February 1927, San Diego, California, amateurs handled all communications while repairs were made to lines washed out by heavy rains. Later in 1927, in the flooded lower Mississippi River valley; in Weeksburg, Kentucky, when a cloudburst hit; and in the New England area, where a tropical storm caused great devastation hams were active in helping. In floods, hurricanes, earthquakes—natural disasters anywhere in the world—the stories of ham participation continue.

Former U.S. President Herbert Hoover, who as Secretary of Commerce after World War I was responsible for amateur radio and took a deep personal interest in it, has written: "The amateurs have performed many signal acts of public service not alone in the field of experiment and research but in the actual trasnsmission of vital messages. Their art has added to the joy of life for literally hundreds of thousands of men, women, boys and girls over the whole nation. Their international communications have a value in bringing a better spirit into the world." Although Mr. Hoover was never a ham himself, his son, Herbert Hoover, Jr., W6ZH, was President of ARRL and long an active ham, as is his son, Herbert Hoover III.

The year 1927 marked a major milestone for amateur radio. It witnessed both the passage of the Radio Act of 1927 (which for the first time gave formal, legal recognition to the Amateur Service) and the convening of the International Radiotelegraph Conference in Washington, in which the United States was forced by the other 70-nation participants to relinquish almost 40 percent of amateur frequency assignments then prevailing. The Radio Act of 1927 provided for the establishment of a Federal Radio Commission to control all radio matters. The creation of a new regulatory agency was actually felt little by amateurs, who continued to function under the continuation of the old Department of Commerce regulations.

The Communications Act of 1934 is the legislation under which amateurs currently operate—although that Act is now in the process of revision. The Federal Communications Commission (FCC) replaced the Federal Radio Commission as the regulatory and licensing agency for the U.S. government, assuming the authority of the Radio Commission and of the Interstate Commerce Commission in matters pertaining to wire telegraph and wire telephone. The definitions applying to amateur radio, however, as well as the structure for licensing and regulating, all remained essentially unchanged. Amateur regulations and the former administrative personnel for amateur affairs were perpetuated by the new agency.

By the mid-1920s amateur radio had evolved its own international language, understood by hams everywhere. It was loosely based on English abbreviations coupled with a mixture of international "Q-Signals" and old telegraphy terms. Thus "very" became "VY," "see you later" became "CUL," "distance" was "DX," "Operator" was "OP," "good morning" was "GM." Telegraphy contributed "73" meaning "best regards" and "88" for "love and kisses." "C" was "YES" and "N" was "NO." "QSO" means "communication," "QSL" means "acknowledgment," and "QTH" means "residence" or, more precisely, "station location."

Rapid Progress

From Paul Godley's 1921 transatlantic amateur contacts to the first two-way transatlantic amateur QSOs in 1923, involving Leon Deloy, 8AB, in France with Fred Schnell, 1MO, and John Reinartz, 1XAM, in Connecticut, to the May 1925 daylight contact on 20 meters between E. J. Simmonds, g2OD, in England and Charles MacLurcan, a2CM, in New South Wales, Australia, the record of operating accomplishments by amateurs was unending. By 1925, in fact, hams had achieved the maximum terrestrial distance—halfway around the globe. In 1961, with the launch of the first amateur-designed and -built space satellite, amateur radio moved into still another era of technical and operating accomplishment.

After eight voyages above the Arctic circle, Captain Donald B. MacMillan and his crew dreaded nothing more than the lengthy isolation for 12 to 14 months. Sensing that radio was the answer and that only amateurs might be able to do the job, MacMillan sought advice from Hiram Percy Maxim at the American Radio Relay League in Hartford. When the MacMillan Arctic Expedition sailed from Wiscasset, Maine, on the auxiliary schooner *Bowdoin* on June 23, 1923, there was a new crew member, Donald H. Mix, 1TS, with a complete 200-meter amateur station donated by the Zenith Radio Corporation. Mix stood regular watches during the months that followed, transmitted weekly reports to the North American Newspaper Alliance, handled expedition message traffic almost daily, and recorded amateur calls heard. Two months later, after crossing Baffin Bay to Cape Sabine, Mix's WNP—"Wireless North Pole"—established a new world's distance record. The little amateur station aboard the *Bowdoin* was an inestimable boon to all aboard. It brought entertainment and news of the world, contacts with home for crew members, messages and reports to the outside world, and a Christmas message from President Coolidge.

The dirigible *Shenandoah* depended on amateur communications with 39 states, Canada, and Mexico during her first nationwide tour. Her later tragic crash has been attributed to neglect of the watchful guardianship of amateur contact.

R.M.S. Tahiti sailed from Sydney to San Francisco in 1924 and returned with a2CM installed in a special radio cabin by its owner, amateur Charles D. MacLurcan, assisted by 16-year-old Jack Davis, a2DS, as operator.

Professor Ernest Merritt of the Department of Physics, Cornell University, wrote in the January 1932 *Proceeding of the Institute of Radio Engineers:*

> Since the amateurs were not allowed to use the longer waves, they went ahead with undiminished enthusiasm to get what results they could with the wavelengths assigned to them. Presumably most of them were not familiar with the theoretical reasons for believing that work with short waves was not likely to prove successful; at any rate, such knowledge of theory as they had did not deter them from trying experiments which the experienced radio engineer would have regarded as foredoomed to failure. When such experiments led to success with 100-meter waves, they tried 50-meter waves and found the results still better. Gradually the wavelength was reduced still further until with a wavelength of about 20 meters it was found possible to

> signal over distances greater than had ever before been reached, and this was only a fraction of the power used by the long-wave stations.

Former ARRL President Herbert Hoover, Jr., W6ZH, wrote these words in his 1964 message commemorating the fiftieth anniversary of ARRL: "They were exciting days in the 1920s when Reinartz, Schnell and Deloy turned the accepted theories of long-distance radio communications upside down, and proved for the first time the enormous usefulness of short wave. This spirit of technical progress and scientific adventure has persisted steadily down through the years and has yielded many solid contributions to the radio art. Each has been a step forward and another feather in the cap of amateur radio."

5

That First Ham License

Is it difficult to become a ham? Is a background in electronics or electricity necessary? Is a technical aptitude required? Not at all.

Consider, for example, a youngster in Iowa. At the age of 5, before he had even entered kindergarten, he was licensed as a Novice, the basic amateur license. In fact, he passed the Morse code portion of the test at age 4! And there is the 78-year-old Wisconsin woman who joined a licensing class in her town and passed the test with flying colors. She had no training or background in anything related to radio, but simply wanted to broaden her interests and activities.

Who knows where that new Novice ticket may lead? Listening to a neighbor's shortwave radio set as a 13-year-old in Philadelphia was the beginning for NBC-TV science correspondent Roy Neal.

"I was fascinated with my neighbor's equipment," Neal recalls, "particularly the glow from the mercury vapor tubes."

That was the start of a 43-year love affair with electronic communication for Neal, K6DUE. He has since covered every U.S. manned space flight, reporting on the Mercury missions directly from Cape Canaveral, and the Gemini and Apollo missions from Houston. He recalls calling on his amateur radio knowledge to improve coverage of the Alan Shepard space flight by linking the Cape and the landing site directly.

A recent survey by the American Radio Relay League revealed that nearly half of today's radio amateurs are or have been citizens' band enthusiasts. They have discovered that the Morse code requirement and the basic radio theory and operating practices one must learn are not at all difficult—and often a great deal of fun!

Newcomers to the ham ranks come from all age groups, educational backgrounds, and occupations. The one quality they share is an interest in discovering a new world, in adding a new dimension to their lives. It is the beginning of an experience that never loses its excitement.

FIG. 5-1 You're never too old—and everyone over 5 isn't too young—to enjoy amateur radio. This youngster operated a Field Day station in Pennsylvania and got some publicity for his hobby in the local newspaper. (Photo courtesy Call-Chronicle Newspapers.)

The Federal Communications Commission, the U.S. licensing and regulatory agency, deliberately made the entry-level license easy to obtain. The Novice license was designed to be a sort of learner's permit to enable the beginner to get on the air quickly and easily, and thereby gain the experience needed to qualify for higher-class licenses.

Novices can use only Morse code, on relatively narrow segments of four of the amateur bands: 10, 15, 40, and 80 meters. But even within those limitations, Novices can and do "talk" around the world and across town, and participate in a variety of on-the-air activities, some just for Novices, but others open to any ham. Many amateurs consider their Novice days the most satisfying of all.

The Novice test is handled through the mail and administered by any amateur holding a General class or higher license who is 18 years of age or older and unrelated to the applicant. The test consists of two parts: a Morse code test and a written quiz. The prospective amateur must demonstrate a code-sending proficiency of 5 words per minute and then answer correctly at least 7 of 10 multiple-choice questions about a 5-minute, plain-language code message at 5 words per minute. Once that part is completed, the volunteer examiner sends the FCC a completed application form, including his or her certification that the code portion has been passed. The FCC then sends back to the examiner the written part, which consists of 20 multiple-choice questions with four-part answers. Fifteen questions must be answered correctly to pass. The questions cover basic radio theory, FCC regulations, and elementary operating practices. There is no cost to the applicant.

Although many hams prepare for the Novice exam on their own, using one of the many available books and a code tape or record, many others prefer to take a class. The American Radio Relay League has a Novice packet entitled *Tune in the World with Ham Radio*, which includes an easy-to-use manual, a student workbook, and a code tape that teaches the Morse code, one letter/numeral at a time. A 5-words-per-minute practice message is also included. Designed for use on a do-it-yourself basis or as a classroom text, this packet has been used successfully by thousands of hams and hundreds of instructors. Preparing on one's own can be satisfying, but it is certainly not the only way.

Today's most popular route to becoming a Novice is by way of a licensing class. These are conducted in cities and towns across the country by local amateur radio clubs, school systems, and individuals. Many use a curriculum and instructors' materials supplied by ARRL. The course usually involves one 2-hour evening class per week for about 10 weeks. At the conclusion, the instructor usually arranges for the Novice test for all participants. For someone starting absolutely from scratch, a licensing class is the easiest way to prepare for the test; it is probably the surest and the most enjoyable. A typical class will have 15 or 20 people of various ages and backgrounds. The cost, if there is any, is nominal. To locate a licensing class in your area, write the ARRL, Newington, Connecticut 06111, for the name and address of a nearby club or instructor.

From the very beginnings of amateur radio, the traditional way to go about getting that first license has been with the help of a ham. He or she may be a friend, a neighbor, or simply someone in your town or area whom you are able to track down because he or she has a ham license plate, works at a local radio or TV station, or is a fellow customer of your local electronics store. Most hams are eager to help newcomers get started. This helping-hand tradition is so well established, in fact, that a name has developed for it—Elmer. A big advantage of having an Elmer is that he or she will assist with setting up a station as well as with Morse code and radio theory. An Elmer will also stick with the beginner after the license comes. For many hams, the relationship with an Elmer is often the beginning of a long-term and special friendship.

Amateur radio is often a family affair. One example is that of a mother and son in upper New York State. She writes:

> My son became interested in amateur radio at the age of 12, which was too young for him to be eligible to enroll in our night school radio class. So I enrolled and he was permitted to attend with me. I was not able to pursue it, but he passed the Novice exam and went on in a few months to pass his General. Now he has a Master's degree in electronic engineering and is employed in antenna research in Florida. When I retired from nursing, I decided that I wanted to get my amateur radio license so I could talk with my son. I received my Novice license without much trouble. I was nearly 70 at the time.

FIG. 5-2 Gerard De Buren, HB9AW, is one of the operators of club station 4U1ITU at the headquarters of the International Telecommunication Union in Geneva, Switzerland. (Photo by Dave Bell, W6AQ.)

A youngster in a New York City suburb was first licensed at 12 and passed his examination for the Extra class license, the highest available to amateurs, at 13½. He has worked amateurs in 38 states and about 25 countries, and says his next project is to launch a 300-foot wire antenna with a balloon!

Learning the Code

The Novice code test includes the letters of the alphabet, numbers (1 through 0) and the most common punctuation marks. There is no one "best" way to learn the code, but there is one basic element in the learning process on which all hams agree: The code should be learned by *sound*, not by sight. You learn what the dits and dahs that make up each letter sound like, not what the dots and dashes look like. The reason is that amateurs always use code by sound; it is never visual. If you learn via the "dots and dashes" route, there is an extra mental step of translation as you hear it on the air. To become really proficient, the visual step must later be unlearned. So it makes good sense to learn the code by sound right from the start.

With practice, the letters and characters cease to be combinations of dits and dahs—each one develops a distinctive sound of its own. For this reason, the best code practice tapes and records send each letter or character itself at a relatively high speed and achieve the slow, 5-words-per-minute rate by means of the space or time between letters. This technique also makes it easier to master the higher speeds needed to upgrade to higher license classes, because it becomes a process of just reducing the time between letters. As many amateurs will testify, this system really works. With surprisingly little practice, the letter *a* quickly becomes *didah* rather than an individual short sound followed by an individual long sound.

Having mastered the letters, numbers, and punctuation marks, the key to proficiency at the Novice level and higher is practice. Regular practice is more productive than is infrequent, intensive practice. For a Novice, that can be a half hour a day for a few weeks. The practice can come from listening to code tapes or records, from listening on the Novice bands on an amateur receiver, listening to code-practice sessions on W1AW (the ARRL headquarters' station—a copy of the schedule is available on request), or by using a code-practice oscillator and key. In the beginning at least, it is a good idea to copy or write down each message as you listen to the code. You will discover that a missing letter or even a word here and there does not prevent getting the entire message.

The Written Test

The written part of the Novice exam consists of a question folder and an answer sheet. There are 20 questions, with four choices for each. There is no time limit and no penalty for guessing. The most effective procedure is to read all the questions first, double-check the instructions, then mark the answers you are sure of. Then go back and choose the best answer of those you aren't quite sure of. You have to answer 15 of the 20 questions correctly to pass the test. Your volunteer instructor will mail the exam back to the FCC for grading.

FIG. 5-3 Although it's possible to learn all you need to know for a Novice license on your own, it's more fun to take a class. This soon-to-be-licensed student attended a licensing class at the Hall of Science, a museum in New York City.

If you passed, you will receive your license—and your own distinctive amateur radio call sign—in the mail a few weeks later. If you did not pass, you will also be notified by mail.

Here are the kinds of questions you can expect:

SAMPLE NOVICE TEST

1) Amateur radio is a radiocommunication service of
   a) expert commercial operators
   b) self-training intercommunication among amateur operators
   c) amateur operators with pecuniary interests
   d) nonlicensed radio transmissions
2) A Novice license is valid for
   a) 1 year and renewable
   b) 5 years and renewable
   c) 1 year and not renewable
   d) 2 years and renewable

3) How can a Novice operate in higher amateur class license bands?
   a) under a control operator
   b) at his own station
   c) with another Novice
   d) on Field Day

4) Novice frequency bands include
   a) 21.1–21.2 and 3.7–3.8 MHz
   b) 7.1–7.15 and 21.1–21.2 MHz
   c) 7150–7200 and 28,100–28,200 kHz
   d) 14.1–14.2 MHz

5) A Novice whose transmitter frequency calibration is accurate to 5 kHz can safely operate how close to 7.1 MHz?
   a) 7150 kHz
   b) 7105 kHz
   c) 7.16 MHz
   d) 7.006 MHz

6) The code group sent as $\overline{\mathrm{AR}}$ means
   a) end of message
   b) end of contact
   c) all right
   d) all ready

7)  is a symbol for
   a) a resistor
   b) an inductor
   c) a transformer
   d) a capacitor

8) The unit of measure for frequency is
   a) ampere
   b) henry
   c) hertz
   d) second

9) Capacitance is measured in
   a) henrys
   b) volts
   c) coulombs
   d) farads

10) Series resistance has a total resistance that is
   a) less than either resistance
   b) the sum of the resistances

c) the difference of the resistances
d) the same as the smallest resistance

11) A Transmatch
a) matches the load to the transmitter
b) attenuates harmonics
c) attenuates parasitics
d) all of these

12) Ohm's law is a relation between
a) power, current, and resistance
b) current, voltage, and power
c) power, current, and voltage
d) current, voltage, and resistance

13) Resistance is measured in
a) henrys
b) farads
c) volts
d) ohms

14) An inductor
a) reacts more with higher frequencies
b) will not pass dc
c) stores electrical energy
d) reacts more with lower frequencies

15) Large-value capacitors and inductors are used in power supplies to
a) act as a tank circuit
b) filter out ac
c) prevent radio-frequency interference
d) tune the power supply

16) An emitter is part of a
a) resistor
b) transistor
c) tube
d) capacitor

17) Transmitter plate voltage is measured between
a) control grid and collector
b) cathode and plate
c) emitter and base
d) screen grid and plate

18) How are signals propagated over great distances?

a) by ground wave
b) sky-wave refraction in the F layers
c) using the skip zone
d) by the mirror effect

19) Line-of-sight communications are a result of which of the above phenomena?

20) Power is measured in
a) hertz
b) amperes
c) volts
d) watts

Operating an Amateur Station

A great deal has been written about the proper way to get on the air and to operate using Morse code. Yet it really boils down to this: The best way is the accepted way. Some of today's operating procedures are mandated by law, but most have evolved out of the early telegraph days and more than half a century of amateur radio. These practices are simply the product of a lot of cumulative experience. They have certainly stood the test of time, and most are just common sense and convenience. Abbreviations and procedural signs, for example, eliminate a lot of unnecessary words and phrases, thereby simplifying communications. They are also universal in all languages, which helps overcome the language barrier.

Some of the most common Morse code abbreviations are listed on the following page. It is not necessary to memorize them all at once; they will become familiar with experience on the air.

The international Q-signals, used as a common means of communication by hams around the world, go back to the earliest days of telegraphy. They are used in two ways: as a question (when followed by a question mark), and as an answer or statement. Some of the most-used Q-signals appear on page 75.

The *RST System* is a means by which one ham reports on the signal of another. It is commonly used as a courtesy near the beginning of an on-the-air contact. A "599" report means that your signal is as good as can be measured on the RST scales. The ratings appear on page 76.

SOME COMMON MORSE CODE ABBREVIATIONS

AA	All after	GND	Ground	SASE	Self-addressed, stamped envelope
AB	All before	GUD	Good	SED	Said
ABT	About	HI	The telegraphic laugh; high	SIG	Signature; signal
ADR	Address	HR	Here, hear	SINE	Operator's personal initials or nickname
AGN	Again	HV	Have	SKED	Schedule
ANT	Antenna	HW	How	SRI	Sorry
BCI	Broadcast interference	LID	A poor operator	SSB	Single sideband
BCL	Broadcast listener	MA, MILS	Milliamperes	SVC	Service; prefix to service message
BK	Break, break me; break in	MSG	Message; prefix to radiogram	T	Zero
BN	All between; been	N	No	TFC	Traffic
BUG	Semi-automatic key	NCS	Net control station	TMW	Tomorrow
B4	Before	ND	Nothing doing	TNX-TKS	Thanks
C	Yes	NIL	Nothing; I have nothing for you	TT	That
CFM	Confirm; I confirm	NM	No more	TU	Thank you
CK	Check	NR	Number	TVI	Television interference
CL	I am closing my station; call	NW	Now; I resume transmission	TX	Transmitter
CLD-CLG	Called; calling	OB	Old boy	TXT	Text
CQ	Calling any station	OC	Old chap	UR-URS	Your; you're; yours
CUD	Could	OM	Old man	VFO	Variable-frequency oscillator
CUL	See you later	OP-OPR	Operator	VY	Very
CW	Continuous wave (i.e., radiotelegraph)	OT	Old timer; old top	WA	Word after
DLD-DLVD	Delivered	PBL	Preamble	WB	Word before
DR	Dear	PSE	Please	WD-WDS	Word; words
DX	Distance, foreign countries	PWR	Power	WKD-WKG	Worked; working
ES	And, &	PX	Press	WL	Well; will
FB	Fine business, excellent	R	Received as transmitted; are	WUD	Would
FM	Frequency modulation	RCD	Received	WX	Weather
GA	Go ahead (or resume sending)	RCVR (RX)	Receiver	XCVR	Transceiver
GB	Good-by	REF	Refer to; referring to; reference	XMTR (TX)	Transmitter
GBA	Give better address	RFI	Radio frequency interference	XTAL	Crystal
GE	Good evening	RIG	Station equipment	XYL (YF)	Wife
GG	Going	RPT	Repeat; I repeat	YL	Young lady
GM	Good morning	RTTY	Radioteletype	73	Best regards
GN	Good night	RX	Receiver	88	Love and kisses

SOME COMMON MORSE CODE Q SIGNALS

QRG	Will you tell me my exact frequency (or that of . . .)? Your exact frequency (or that of . . .) is . . . kHz.
QRH	Does my frequency vary? Your frequency varies.
QRI	How is the tone of my transmission? The tone of your transmission is . . . (1. Good; 2. Variable; 3. Bad).
QRK	What is the intelligibility of my signals (or those of . . .)? The intelligibility of your signals (or those of . . .) is . . . (1. Bad; 2. Poor; 3. Fair; 4. Good; 5. Excellent).
QRL	Are you busy? I am busy (or I am busy with . . .). Please do not interfere.
QRM	Is my transmission being interfered with? Your transmission is being interfered with . . . (1. Nil; 2. Slightly; 3. Moderately; 4. Severely; 5. Extremely.)
QRN	Are you troubled by static? I am troubled by static . . . (1–5 as under QRM).
QRO	Shall I increase power? Increase power.
QRP	Shall I decrease power? Decrease power.
QRQ	Shall I send faster? Send faster (. . . wpm).
QRS	Shall I send more slowly? Send more slowly (. . . wpm).
QRT	Shall I stop sending? Stop sending
QRU	Have you anything for me? I have nothing for you.
QRV	Are you ready? I am ready.
QRW	Shall I inform . . . that you are calling him on . . . kHz? Please inform . . . that I am calling on . . . kHz.
QRX	When will you call me again? I will call you again at . . . hours (on . . . kHz).
QRY	What is my turn? Your turn is numbered . . .
QRZ	Who is calling me? You are being called by . . . (on . . . kHz).
QSA	What is the strength of my signals (or those of . . .)? The strength of your signals (or those of . . .) is . . . (1. Scarcely perceptible; 2. Weak; 3. Fairly good; 4. Good; 5. Very good).
QSB	Are my signals fading? Your signals are fading.
QSD	Are my signals mutilated? Your signals are mutilated.
QSG	Shall I send . . . messages at a time? Send . . . messages at a time.
QSK	Can you hear me between your signals and if so can I break in on your transmission? I can hear you between my signals; break in on my transmission.
QSL	Can you acknowledge receipt? I am acknowledging receipt.
QSM	Shall I repeat the last message which I sent you, or some previous message? Repeat the last message which you sent me [or message(s) number(s) . . .].
QSN	Did you hear me (or . . .) on . . . kHz? I did hear you (or . . .) on . . . kHz.
QSO	Can you communicate with . . . direct or by relay? I can communicate with . . . direct (or by relay through . . .).
QSP	Will you relay to . . .? I will relay to . . .
QST	General call preceding a message addressed to all amateurs and ARRL members. This is in effect "CQ ARRL."
QSU	Shall I send or reply on this frequency (or on . . . kHz)? Send or reply on this frequency (or on . . . kHz).
QSV	Shall I send a series of Vs on this frequency (or . . . kHz)? Send a series of Vs on this frequency (or . . . kHz).
QSW	Will you send on this frequency (or on . . . kHz)? I am going to send on this frequency (or on . . . kHz).
QSX	Will you listen to . . . on . . . kHz? I am listening to . . . on . . . kHz.
QSY	Shall I change to transmission on another frequency? Change to transmission on another frequency (or on . . . kHz).
QSZ	Shall I send each word or group more than once? Send each word or group twice (or . . . times).
QTA	Shall I cancel message number . . .? Cancel message number . . .
QTB	Do you agree with my counting of words? I do not agree with your counting of words. I will repeat the first letter or digit of each word or group.
QTC	How many messages have you to send? I have . . . messages for you (or for . . .).
QTH	What is your location? My location is . . .
QTR	What is the correct time? The time is . . .

READABILITY

1—Unreadable.
2—Barely readable, occasional words distinguishable.
3—Readable with considerable difficulty.
4—Readable with practically no difficulty.
5—Perfectly readable.

SIGNAL STRENGTH

1—Faint signals, barely perceptible.
2—Very weak signals.
3—Weak signals.
4—Fair signals.
5—Fairly good signals.
6—Good signals.
7—Moderately strong signals.
8—Strong signals.
9—Extremely strong signals.

TONE

1—Sixty-cycle ac or less, very rough and broad.
2—Very rough ac, very harsh and broad.
3—Rough ac tone, rectified but not filtered.
4—Rough note, some trace of filtering.
5—Filtered rectified ac but strongly ripple-modulated.
6—Filtered tone, definite trace of ripple modulation.
7—Near pure tone, trace of ripple modulation.
8—Near perfect tone, slight trace of modulation.
9—Perfect tone, no trace of ripple or modulation of any kind.

Unlike a telephone conversation, it is not always possible for amateurs to transmit and receive simultaneously. To make it easier to change from transmitting to receiving, hams have adopted a set of standard signals that are used at the end of a transmission. Each of these signals has a Morse code and a voice counterpart. The line over a signal means it is sent as a single character rather than as two separate letters.

MORSE CODE AND VOICE ENDING SIGNALS

Voice	*Code*	*Meaning*
over	$\overline{\text{AR}}$	after call to specific station
wait, stand by	$\overline{\text{AS}}$	self-explanatory
roger	R	all received correctly
go	K	any station transmit
go only	$\overline{\text{KN}}$	addressed station only transmit
clear	$\overline{\text{SK}}$	end of contact
closing station	CL	going off the air

A First On-the-Air Contact

The first contact is an experience that no amateur ever forgets. Even a later exotic foreign contact will not surpass it. The first time you hear someone answering your call, with your station's call sign, is an experience that is hard to beat.

A first contact (QSO) may go something like this (the newly licensed amateur is KB1XYZ):

CQ CQ CQ DE KB1XYZ KB1XYZ

An impatient wait to listen for a response. Nothing. Try again.

CQ CQ CQ DE KB1XYZ KB1XYZ KB1XYZ

After another brief wait, suddenly, there it is.

KB1XYZ KB1XYZ KB1XYZ DE W2HD W2HD $\overline{\text{AR}}$

It's your call sign over the air for the first time. Someone has heard you! W2HD is acknowledging your call, indicating he or she is ready for a contact, and telling you to go ahead with your next message.

W2HD DE KB1XYZ $\overline{\text{BT}}$ RST 589 WORCESTER MASS $\overline{\text{BT}}$ NAME BILL $\overline{\text{BT}}$ GM W2HD DE KB1XYZ $\overline{\text{KN}}$

You reply, giving W2HD a reading on his or her signal at your location, state that your name is Bill, and wish him or her a good morning. Your $\overline{\text{KN}}$ ending lets W2HD—and everyone else listening—know that you've ended that particular transmission and are waiting for W2HD, and no one else, to get back to you. Since you are both already in contact, you use call signs only once every 10 minutes.

> KB1XYZ DE W2HD R $\overline{\text{BT}}$ RST 678 READING PA $\overline{\text{BT}}$ NAME HARRY $\overline{\text{BT}}$ GLD MT U $\overline{\text{BT}}$ HW LNG U BN HAM?

W2HD tells you he has received your entire message, gives you a report on your signal at his location, tells you his name is Harry, says he is glad to meet you, and asks how long you have been a ham. He assumes you are recently licensed because you are operating in a Novice band. The question mark lets you know he has turned it back to you.

> ABT 2 MOS $\overline{\text{BT}}$ UR MY FIRST CONTACT

Now that contact is clearly established, it is no longer necessary to repeat call signs or to use an ending sign. When you have answered the question, your silence indicates that you are through sending.

> WELCOME ABOARD $\overline{\text{BT}}$ PLSD BE UR FIRST QSO $\overline{\text{BT}}$ HOPE U NJY HAM RADIO AS MCH AS I DO

It is back to you again, so you reply

> TNX FR COMING BACK TO ME OM $\overline{\text{BT}}$ IM A BIT NERVOUS $\overline{\text{KN}}$

W2HD responds

> UR DOING FINE $\overline{\text{BT}}$ STICK WITH IT $\overline{\text{BT}}$ UR GG TO HV GD FIST $\overline{\text{KN}}$

When the contact has ended and you are ready to sign clear, you might send:

> MNY TX QSO HARRY $\overline{\text{BT}}$ 73 $\overline{\text{SK}}$ W2HD DE KB1XYZ

You have now thanked W2HD, wished him "best regards," and indicated that this contact is ended and you are open for others. If you were going off the air and shutting down your station, you would have sent CL instead of $\overline{\text{SK}}$.

The FCC rules no longer require that each contact be recorded in a log, but most hams still maintain a complete log of their contacts. For one thing, that record is useful in exchanging QSL cards, the postcard-size acknowledgment of a contact. A number of standard logbooks are available, but there is nothing wrong with making your own.

The FCC does require that some sort of log be kept at every U.S. amateur station. It must contain the station call sign and location and the licensee's signature. Entries must include the date the station began operation, the date operation is ended, and the time and date of any operation under control of another licensed amateur (who must sign the logbook next to the entry). The only other required record is of any *third-party traffic*—messages sent for someone else. A typical logbook provides space for the date and time of the contact, the call sign of the other station, both signal reports, the frequency on which the contact took place, the time it ended, comments, and whether a QSL card was sent or received.

Because of the variation in international time zones, hams have adopted a system known as Universal Coordinated Time (UTC). Formerly known as Greenwich Mean Time (GMT), it is the time at the zero meridian. Using a table, it is easy to convert the time anywhere in the world to UTC. This 24-hour system, which specifies time from 0000 to 2400, eliminates confusion about whether a contact was at 10 A.M. or 10 P.M. For convenience, many amateurs have a 24-hour clock in their station, set to UTC.

Some Facts About Time Conversion

The chart below has been arranged to show time zones, used by most amateurs in the North American continent and Universal Coordinated Time, used universally as a standard. The advantage of UTC is that it is the universally understood reference throughout the world. ARRL recommends that all amateur logging be done in UTC.

Time zone letters may be used to identify the kind of time being used. For example, UTC is designated by the letter Z.

In converting from one time to another, be sure the day or date corresponds to the new time. That is, 2100 EST on January 1 would

UTC	*EDT/AST*	*CDT/EST*	*MDT/CST*	*PDT/MST*	*PST*
0000*	2000	1900	1800	1700	1600
0100	2100	2000	1900	1800	1700
0200	2200	2100	2000	1900	1800
0300	2300	2200	2100	2000	1900
0400	0000*	2300	2200	2100	2000
0500	0100	0000*	2300	2200	2100
0600	0200	0100	0000*	2300	2200
0700	0300	0200	0100	0000*	2300
0800	0400	0300	0200	0100	0000*
0900	0500	0400	0300	0200	0100
1000	0600	0500	0400	0300	0200
1100	0700	0600	0500	0400	0300
1200	0800	0700	0600	0500	0400
1300	0900	0800	0700	0600	0500
1400	1000	0900	0800	0700	0600
1500	1100	1000	0900	0800	0700
1600	1200	1100	1000	0900	0800
1700	1300	1200	1100	1000	0900
1800	1400	1300	1200	1100	1000
1900	1500	1400	1300	1200	1100
2000	1600	1500	1400	1300	1200
2100	1700	1600	1500	1400	1300
2200	1800	1700	1600	1500	1400
2300	1900	1800	1700	1600	1500
2400*	2000	1900	1800	1700	1600

*0000 and 2400 are interchangeable

be 0200Z (UTC) on January 2; similarly, 0400Z on January 2 would be 2000 PST on January 1.

A good method is to use UTC (Z) for *all* amateur logging, schedule making, QSLing, and other amateur work. Otherwise, confusion, with all the different time zones, is inevitable. Leave your clock on UTC.

The exchange of *QSL cards* is a traditional courtesy that had its origins in the early days of amateur radio. It is difficult to conceive of an active ham not having them, although there is no regulation requiring that they be sent. Varying from the plain, straightfoward variety to the exotic or elaborate, QSL cards are highly personal

FIG. 5-4 QSL cards abound in ham shacks around the world. This wallfull of cards appeared at a Sister Cities convention in Phoenix, where delegates from a dozen countries learned how amateur radio can bridge the gap between the world's peoples.

statements about you, the station operator. Standard printed cards are available from printers who advertise in amateur publications. But those who prefer more personalized cards will have them custom made by a printer or will make them themselves. Basically, a QSL card displays its owner's call sign prominently, together with the name, address, and space for recording the details of the contact: date, time, frequency, and signal report. It may also list principal station equipment and membership in amateur organizations.

For some amateurs, the exchange of QSL cards is accomplished individually, one at a time. For ordinary purposes the postage is not a significant burden. Active DXers, on the other hand, can have a problem, as the cost for sending many cards overseas can be enormous. The solution is the *QSL bureau.* Established and operated by individual volunteers and organizations such as ARRL, they simplify the process of sending and receiving QSL cards. Outgoing cards are sent to the ARRL Overseas QSL Service, which sorts them and sends them in bulk to the countries of the world. There, incoming QSL bureaus forward them to the addressee.

To get cards via the incoming QSL bureau serving your area, find the address in *QST* or other amateur magazine and keep several self-addressed envelopes on file there. In addition, tell your overseas contacts to "QSL via the bureau," and they will send their cards to the right place.

As you spend more and more time operating your amateur station, you will gain the confidence needed to become highly skilled in the procedures. You'll be surprised at how quickly you feel comfortable at the key, perhaps even becoming someone else's Elmer before too long.

6

Setting Up a Station

An amateur radio station consists basically of just three elements: a transmitter, a receiver, and an antenna. All three can be combined in a single, hand-held unit, a compact under-dash automobile rig with a whip antenna on the rear deck, or an array of components that nearly fills a small room. A typical Novice station will fit readily on a small desk or table.

But how does one go about assembling that first station? Where does one start—with what—and where? One newly licensed ham's experience provides some answers:

> There are companies which manufacture kits you can put together yourself, making them less expensive than buying the gear ready-built. I helped a ham friend build a five-band transceiver. I learned what all kinds of components looked like when I checked off the contents of the kit with the list of what was

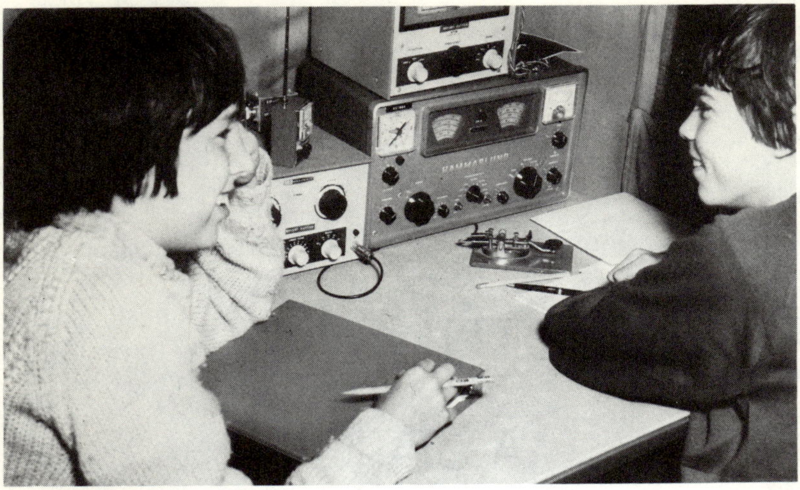

FIG. 6-1 Just a corner of a room affords space enough for equipment capable of working the world. (Photo by Christopher Wiles.)

> supposed to be in it. After much practice, I learned to solder pretty well. It took several evenings and the next to the last thing we did was to go over every step to make sure we had done just what we were supposed to do. The last act was to hold our breath as we plugged it in, hooked up the antenna and turned the dial. It worked! What a feeling of accomplishment!

In the early days of amateur radio, that feeling of accomplishment was part of the inaugural process for every ham, because building one's own gear was the only way to get started. Now, of course, the newcomer has access to a wide variety of commercially manufactured equipment and can get on the air without building his or her rig. This approach does, however, eliminate a lot of fun and satisfaction—not to mention the opportunity to save some money. And the new ham will never have quite the understanding of what is inside that box and how it works that firsthand construction experience will provide. Modern kits are a good compromise. They are less expensive than completely finished gear, and they are reasonably foolproof. No special skill or technical training is needed, either.

There is no question that the biggest asset a new Novice can

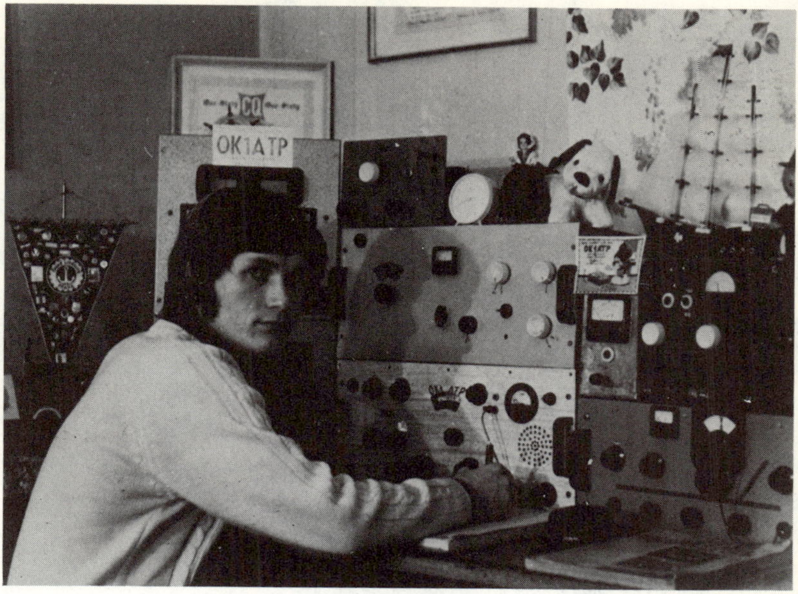

FIG. 6-2 OK1ATP, a Czech amateur, built this ocean-spanning station with his own hands.

have in setting up a station is an Elmer, a helping ham. An Elmer can help cut through all the mystery and confusion of selecting gear from the bewildering array that is available. He or she can help relate equipment needs to the Novice's budget and particular areas of interest in amateur radio. Elmers can assess the relative merits of types of equipment, design features, and operating conveniences. And perhaps most important of all, Elmers can help locate sources of equipment, especially secondhand gear that may be available from other hams.

If possible, it is a good idea to get some experience with amateur equipment of various types and brands before making a personal investment. This has a number of advantages: New Novices can decide for themselves what features are particularly useful or desirable, and what equipment they feel comfortable or uncomfortable with—the type best suited to their needs, interests, and requirements. An Elmer's station, stations belonging to other nearby hams,

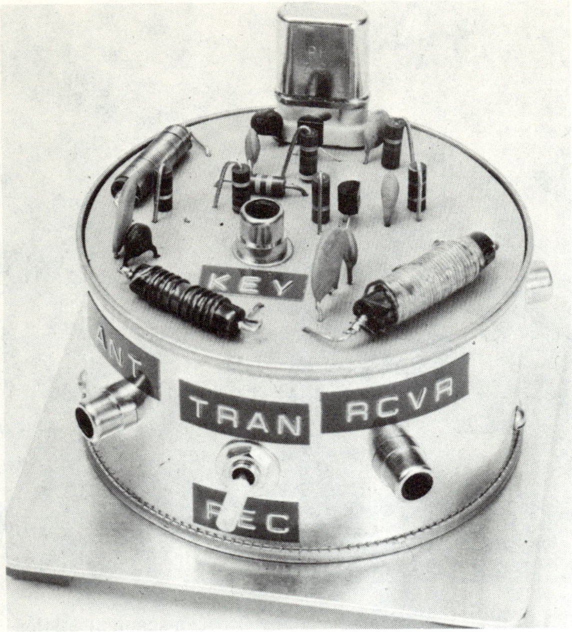

FIG. 6-3 No larger than a fist, this low-power code transmitter is entirely homemade—in a tunafish can!

and club or school stations may provide operating opportunities. One thing the newcomer may be sure of is that no two stations will be exactly alike; and each owner will be quite prepared to extol the virtues of his or her particular setup!

Try Used Equipment First

For sheer economy it is hard to beat some of the buys that are frequently available in used amateur equipment. A recently licensed ham cites an example: "A fellow I work with had some old ham radio junk in his attic he wanted to sell. So I bought the whole works for $50. I got a World Radio Laboratory transceiver, a power supply, a Heath RF generator, a 5-inch oscilloscope, a voltmeter and a transistor checker."

FIG. 6-4 A consultation with an Elmer—another mystery solved. Experienced hams can help with almost any problem confronting the new Novice. (Photo by Joel Kleinman, WA1ZUY.)

All hams accumulate excess equipment. They may trade it for items they need, they may find buyers through local radio clubs or contacts with new amateurs, or they may take it to a popular ham event known as a flea market. These are held in conjunction with ham gatherings of all kinds in just about every community where hams live. A flea market can be a treasure trove of great bargains for a new Novice, especially if an Elmer helps avoid pitfalls. Another excellent source of used ham gear is dealers who take equipment in trade on new products. For the Novice who is not fussy about having the latest gadgetry, a complete and perfectly workable station can be had for less than $200.

Everything considered, the least expensive, least complicated, and simplest-to-operate equipment is probably a Novice-type transceiver or transmitter–receiver combination for CW (code) operation only. Most such rigs will cover all four of the bands on which Novices are permitted to operate. With such a station, the new ham will have a relatively modest investment; more sophisticated equipment can come as experience and burgeoning interests require it. Outgrown

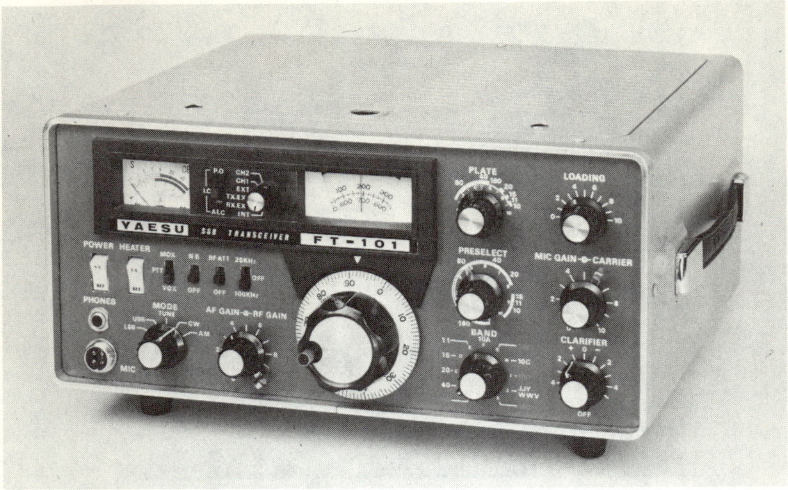

FIG. 6-5 A transceiver combines receiver and transmitter in one unit. Only antenna, key, SWR meter—and operator—need be added.

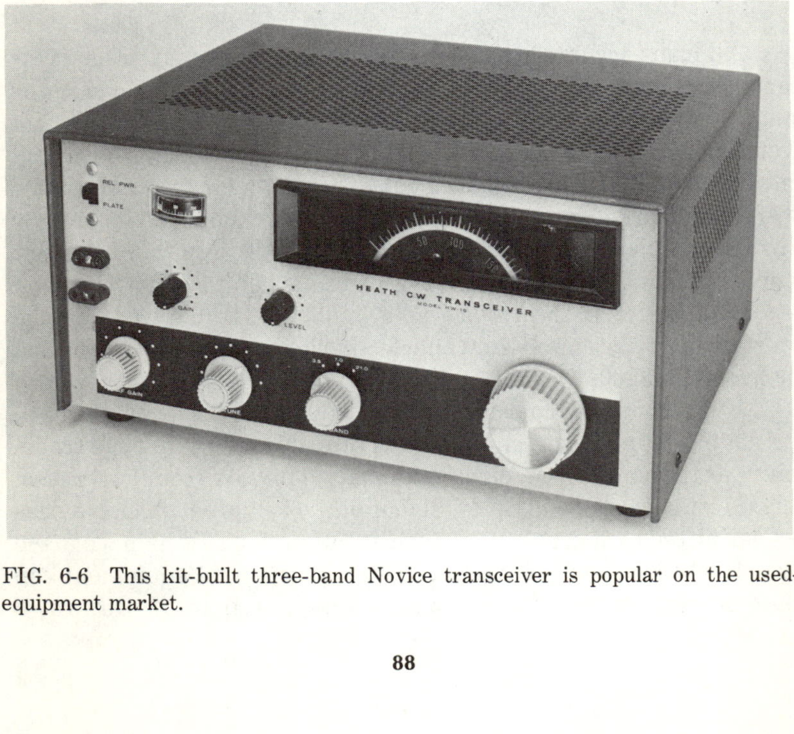

FIG. 6-6 This kit-built three-band Novice transceiver is popular on the used-equipment market.

gear will have a ready trade-in or resale value when the time comes. Novice rigs are available new, used, and in kit form, so there is a range of choice.

If, on the other hand, the new ham wishes to start with at least some equipment suitable for more advanced use later, many Elmers will suggest that a good receiver should be the first choice. A receiver is less likely to be outgrown sooner than a transmitter and is generally more versatile in its applications to different kinds of operating. Another experienced ham can help check a receiver for *stability*, to ensure that its tuning does not drift once set by the operator; for *selectivity*, the ability to reject unwanted signals; for *sensitivity*, the ability to pick up weak signals; and for accurate calibration, or *dial accuracy*.

A good telegraph key is an important requirement for the new ham. "Bugs" and electronic keyers, devices that make high-speed sending easier, are popular with many experienced CW operators. The Novice, however, should stick to the basic telegraph key, or "straight key," until his or her code speed reaches the 15-words-per-minute level. Although the key is just a device to turn a transmitter on and off to form code characters, any straight key worth purchasing should feature ease and latitude in two adjustments. One of these is *spring tension*—how much force the operator must apply to close the key and turn on the transmitter. The other is *contact distance*—the distance between the two metal points, or contacts, that come together when the operator presses down on the key. The

FIG. 6-7 Every Novice is proud to send crisp Morse code with the basic straight key.

best straight keys do not combine these two adjustments into one. There are many excellent keys on the market; the things to look for are a heavy, substantial base, positive closure of the contacts, and those important adjustments for contact distance and spring tension. Many operators form a strong sentimental attachment to that first key—especially if it is a good one!

Getting Ready

Receiver, transmitter, and key should be placed atop a table or similar surface. An old desk or a sturdy table will serve well if it is deep enough to afford writing space as well as room for the rig and its accessories. A comfortable chair and a lamp complete the basics.

The receiver is usually the piece of equipment requiring the most adjustment and tuning during on-the-air operation, so the arrangement of the station should be built around the receiver if the station receiver and transmitter are not one unit. The transmitter, next in importance, should also, of course, be convenient to reach. Placement of the key is next; there should be sufficient room so that the operator's entire forearm can be supported on the tabletop. The key should be well clear of other objects that could interfere with arm movement in reaching other equipment. The key should be bolted to the table top or heavily weighted so that it will not shift during operation—nothing can be more aggravating in the middle of a QSO!

Location of the station itself is probably beyond the new ham's control; it will be dictated by available space and the wishes of other members of the family. To the extent that a choice is possible, however, it should be out of everyone's way and away from family traffic, to avoid distraction when operating and to reduce opportunities for inexperienced or unqualified experimentation with the equipment. There should be direct access to a power source or convenience outlet, preferably to a circuit with little or nothing else connected to it. In particular, it should be a circuit not used by appliances such as toasters, coffee makers, bathroom heaters, or other heavy-current users. An outlet capable of supplying as little as 10 amperes may be adequate for a station; 20 amperes should be more than enough for any ham gear. There should be provision for a master

FIG. 6-8 Receiver, transmitter, clock, key, accessories—all are arrayed for smoother operation and maximum visibility in the station of this New Jersey ham.

switch so that all power to the station can be turned off when the station is not in use.

Short of a separate room for the ham shack (a rare ham luxury!), the most common locations are a den, a corner of a bedroom, the basement, or the attic. Each of these has advantages and disadvantages. One well-known and very active amateur of many years has an entirely adequate station in a closet! In addition to power, the station requires access to a ground and to the antenna. Good grounding is essential; it can be accomplished by a connection to a metal cold-water pipe if one directly connected to an underground water-distribution system is accessible, or to a pipe or rod driven 4 to 6 feet into the ground. In many stations, the connection between the station and the antenna consists of a length of coaxial cable, a cable slightly heavier than an outdoor extension cord. As a general rule, the shorter the length of this cable, the more efficiently the transmitter's power will be carried to the antenna since, as with any transmission line, there is an energy loss. These are considerations that may affect the choice of a location for the station.

Antenna Placement Is Vital

As noted earlier, it is the antenna that primarily determines how "big" a signal the station is going to put out. And the key factor is height—the higher it is, the better. The really dedicated ham will always seek a home site that is on top of a hill or at least above surrounding territory. To the uninitiated, that massive structure in a DXer's or contester's back yard is an "antenna." Not so. The antenna is actually just the array at the top of the structure; the rest is a tower to get it up high.

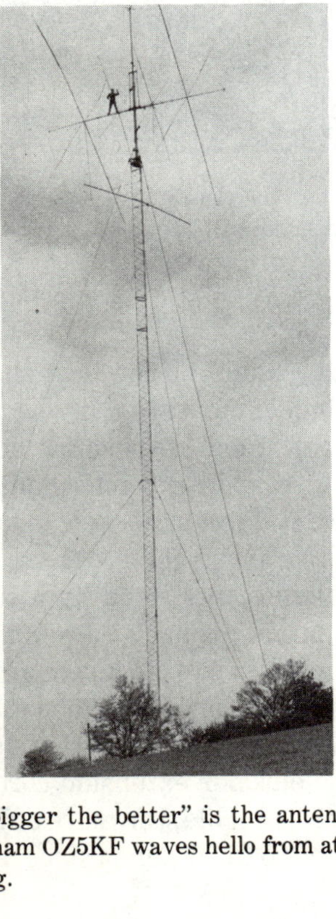

FIG. 6-9 "The bigger the better" is the antenna watchword of the amateur. Danish ham OZ5KF waves hello from atop his contribution to the best of the big.

On the other hand, a city apartment dweller is apt to be severely restricted in antenna options. He or she may have to settle for an unobtrusive wire out the window or even around the ceiling in several rooms. Such an arrangement can work very satisfactorily, as many a ham has discovered. A little ingenuity can make a lot of difference, and knowledge of the principles of antenna design can be a big help in such a situation.

Hams have their equivalent of old-time barn-raising parties, but they call it antenna-raising. They enjoy pitching in when a fellow club member, a ham neighbor, or especially a new ham has an antenna to put up. A collective effort can make short work of the project. One ham, for example, gave this report on a Field Day setup: "Last year it took us 23 minutes to get a rig on the air. This year we planned the setup carefully and packed the truck in just the right way. Had a rig on the air in just nine minutes flat."

For a Novice putting up a first antenna, an antenna party is not only an excellent way to tap the expertise of knowledgeable hams, but also to have fun and get to know area amateurs. Often just a

FIG. 6-10 In keeping with ham radio's long tradition of friendship and spirit, everyone pitches in when a new amateur in the neighborhood has an antenna to raise.

casual mention at a club meeting that someone has an antenna to be raised is all that is necessary to get the ball rolling and turn out the volunteers. Each of the three most popular antennas for beginners—the dipole, the vertical, and random-length wire—is relatively simple to erect and put into operation, and all are inexpensive.

The half-wave dipole is probably at the top of the list for most Novices because it is the easiest and the most forgiving—it will work just about anywhere with just about any kind of installation. It is a single-band antenna but can be adapted for use on more than one band, even all four Novice bands, fairly simply.

There are some practical cautions to be observed in selecting an antenna location. The antenna should not cross over or under a power line, because contact with the line in the wind or during a storm could cause electrocution. It should not run parallel and close to a power line because it will receive a lot of electrical noise, an annoying type of interference to reception, from the line. Similarly, large metal surfaces—gutters, beams, or metal siding—should be avoided as much as possible because they can reduce signal radiation by absorbing radiated energy.

The components of a dipole antenna are simple: two lengths of wire, each equal to one quarter-wavelength at the transmitter frequency, three insulators, wire and heavy hooks or other means of attachment to supports, a length of coaxial cable, and a connector for the transceiver or transmit/receive switch in the station.

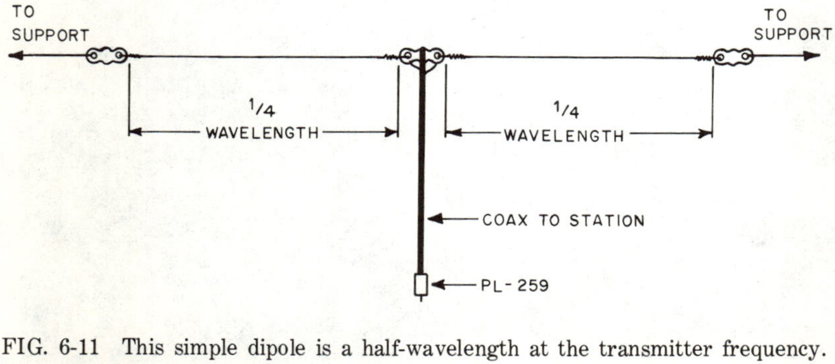

FIG. 6-11 This simple dipole is a half-wavelength at the transmitter frequency. Depending on whether end or center supports are available, one or both ends may be dropped, resulting in a "sloper" or "inverted V."

The total length of the two pieces of wire in the antenna can be calculated using the equation

$$\text{antenna length (feet)} = \frac{468}{\text{frequency (MHz)}}$$

Each of the two pieces of wire should be cut a little longer than a quarter-wavelength to allow for attachment to the insulators at each end and in the middle. The preferred wire for antennas is heavy-gauge Copperweld, a wire of steel core covered with a copper jacket for both conductivity and strength. Ideally, the entire antenna should be as high as possible. If, however, both ends cannot be at the same height, one or even both can be lower if the middle is sufficiently elevated. With one end lower than the other, the dipole is known as a *sloper*. With both ends dropped, it becomes an *inverted V*. With an inverted V, the ends should be as far apart as possible. With either an inverted V or a sloper, the ends of the antenna must be well out of reach of anyone on the ground because they may carry enough voltage for a nasty burn. Rope or another insulating material, such as

FIG. 6-12 As this ham in Illinois is discovering, a sturdy tree is a fine support for the end or center of a dipole.

fishing line, can be used to connect the insulators at the ends of the antenna to the ground, trees, poles, or other supports.

The antenna is connected to the station with a sufficient length of coaxial cable to reach the antenna after it is elevated into position. The coax best suited for Novice use is identified as RG-58/U (it might also be labeled /AU, /BU, or Polyfoam). This type of coax is about ¼ inch in diameter, reasonably light in weight, and flexible. The coax is connected at the center insulator. Carefully cut away a few inches of the outer covering from one end of the coax; underneath is a layer of copper braid, inside of which is another insulating layer, and at the center the other conductor. Twist the strands of copper braid together to form a single wire; connect that to the antenna wire on one side of the center insulator and the other conductor to the other side of the antenna. Then tape or otherwise support the coax itself to the center insulator and solder the two connections. That done, the entire antenna is ready to be raised into position.

The other end of the coax is attached to the station with a radio-frequency (RF) connector, the simplest of which (PL-259 is a common one) is available in solderless form. In either case, the dealer

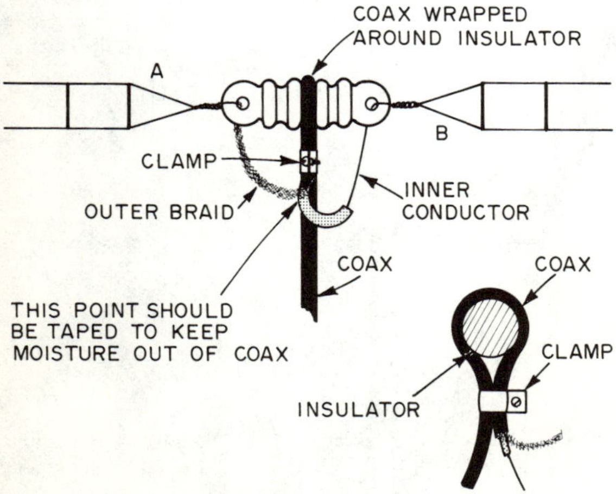

FIG. 6-13 The critical coaxial connection to a dipole. Wrapping it with electrical tape keeps weather out and cable quality high.

from whom it is purchased can explain how to attach it. For a transceiver, the connection is directly into the antenna jack at the rear of the unit. For a separate transmitter–receiver setup, a TR (transmit-receive) switch must be used so that the antenna can serve both units. It switches the antenna between receiver and transmitter so that only one of them is connected to the antenna at a time. A TR switch may be built into a transmitter or transceiver. If not, it is available at the same retail stores selling coax and antenna wire. The switch is connected to the transmitter and the receiver with short lengths of coax and RF connectors.

To check on how well the antenna is actually tuned to the frequency for which it will be used, a device called a VSWR bridge—or SWR meter, as it is commonly known—is required. Most hams have one. It measures the voltage standing-wave ratio. The lower the ratio, the better the antenna is tuned to the transmitter; the reading should be 2:1 or less. The SWR meter is connected into the coax line between the transmitter and antenna. In use, a control knob on the SWR meter is adjusted until a dial pointer is set at an indicated mark. If the SWR meter does not have its own instructions for use, an experienced ham can help. If the SWR is greater than 2:1, the

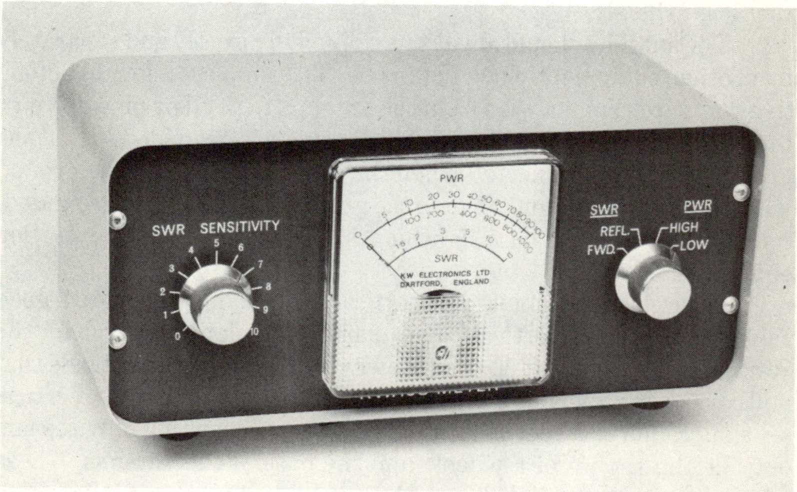

FIG. 6-14 The SWR meter is a must for keeping antenna system efficiency high.

antenna wire should be shortened or lengthened to bring it into proper tune with the transmitter frequency.

The dipole antenna will normally operate on just one band. There are two ways in which it can be adapted to accommodate another band, even all four Novice bands. The simplest is by means of an antenna tuner or Transmatch. This is a device that electrically adjusts the "length" of the antenna to match various wavelengths. It, too, is connected into the coax line between the transmitter and the antenna.

The second way to "add bands" to a dipole antenna is to cut additional pairs of antenna wires for each band desired. Each pair can then be connected and soldered to the coax line at the center insulator. With this multiband installation, the transmitter will "automatically" select the antenna for the band to which it is tuned; the others will have no effect. The antenna wires should all be attached to supports so that none touch or cross. There will, of course, be a considerable difference in their lengths. If you decide to use this type of setup, take care that your antenna does not radiate signals on more than one band at once. This is especially likely to occur on the 80-meter band. To avoid causing interference for others, have your antenna system checked out by an experienced amateur before putting it on the air.

The antenna should always be grounded directly and separately from other equipment when not in use. The simplest and most effective way is to clamp a wire from the antenna to a rod or pipe driven 4 to 6 feet into the ground. A switch will have to be provided so that the ground can be disconnected when the antenna is in use. The ground wire should run straight down to the ground rod. Properly grounded, the antenna will not be a lightning hazard to the home or station equipment.

Where space is limited, a vertical antenna may provide a good solution. A full-size vertical is one quarter-wavelength, which for the 80-meter band is about 60 feet. However, 20-foot-long verticals that will cover all the Novice bands are available commercially. A vertical antenna requires a special grounding system consisting of wires laid out like the spokes of a wheel from the base of the antenna. These wires, known as "radials," must be cut to one-quarter of the wavelength at the frequency to be used. They may be laid on the surface of the ground but are usually buried (no deeper than a few inches) to keep them out of the way and to prevent their being damaged. The greater the number of radials, the better the antenna will perform. As few as two radials per band can be used, however. Instructions for attaching the radials and the coax feed line to the antenna are usually provided by the manufacturer. RG-58 coax with PL-259 connectors serve as well for feed line to a vertical as they do for a dipole antenna.

In situations where neither a dipole nor a vertical is feasible, the random-length wire antenna is a very satisfactory alternative. Although no specific length is required, the wire should be as long as possible, which may be and often is just long enough to run along the ceilings of a few apartment rooms. The matching of the electrical length of a random-length wire antenna to each band is accomplished with an antenna tuner or Transmatch connected into the coax line between the antenna and the transmitter. Antenna tuners are available at a range of prices; units to cover the four Novice bands can be quite inexpensive.

No amateur should ever forget that there are hazards in main-

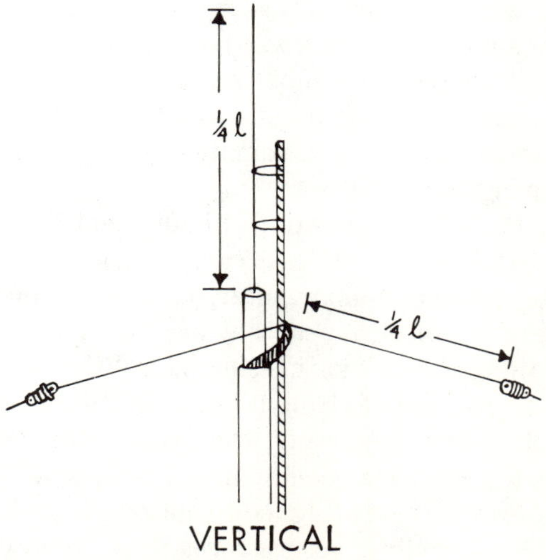

FIG. 6-15 A vertical antenna is ideal when space limitation make a dipole inconvenient. This type of antenna can operate on one or several bands.

taining a radio station. All hams deal with very high voltages in equipment and with height while installing antennas. Safety is a state of mind that should become a habit right from the start.

Accessories Make Operating More Enjoyable

An *SWR bridge* and an *antenna tuner* are two useful accessories for station operation that have already been described. There are a great many others a ham can build or otherwise acquire to enhance his or her enjoyment of amateur radio or even to satisfy a passion for gadgetry. They can represent an investment limited only by the ham's own budget.

There are several accessories that most amateurs sooner or later find desirable for convenient operation. One, for example, is called a *dummy load.* When tuning a transmitter, an on-the-air signal can cause interference to other amateurs. Yet there must be some kind of signal outlet or the transmitter can be damaged. A dummy load, which is essentially a large resistor, supplies that outlet in place of the antenna. It allows the transmitter to be tuned, adjusted, or band-switched without a signal going out over the air. It is usually connected with an antenna switch that can handle several different antennas as well as a dummy load; this is more convenient than unscrewing connections.

The *multimeter,* a piece of test equipment found in many ham shacks, is used to measure voltage, current, and resistance. The most popular is the volt-ohm-milliammeter, or VOM for short. VOMs come in many sizes and price ranges; they are handy for use almost anywhere. The things to look for in selecting a VOM are the number of ranges it has, the ease of changing ranges, its readability, and the ohms-per-volt rating. Generally, the greater the rating, the more useful the meter will be; a good-quality VOM provides 20,000 ohms/volt. The VOM should be designed to operate in an RF field; otherwise, it can give wildly inaccurate readings when used while the transmitter is operating.

Another popular multimeter is the vacuum-tube voltmeter, or VTVM. It is a bulkier, plug-in unit that is often more accurate than a

FIG. 6-16 A dummy load replaces the antenna but does not radiate a signal—use it to avoid interfering with other on-the-air contacts when making band changes and transmitter adjustments.

VOM. A newer type of multimeter is the battery-powered FET VOM. It is a solid-state device that combines the sensitivity of a VTVM with the portability of a VOM—at a higher price. It is especially important to read the instructions carefully before using a multimeter for the first time because carelessness can damage the meter or result in a dangerous shock.

FCC regulations no longer require that every contact be recorded in a station log. Yet most hams find it useful to keep a complete log. A log helps in checking back on previous QSOs, for contest qualification, exchanging QSL cards, and more. A 24-hour station clock is a great convenience, not only for logging, but because

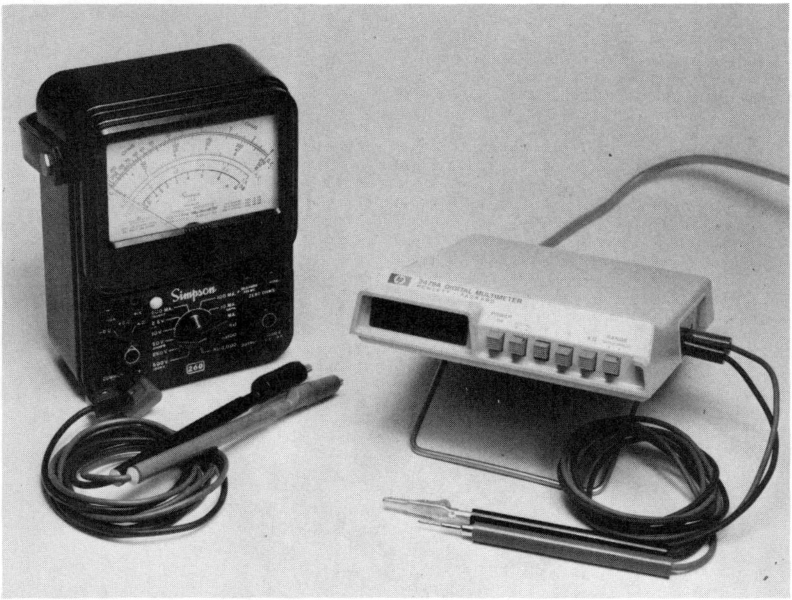

FIG. 6-17 The volt-ohm-milliameter, or VOM, is the ham's basic test instrument.

it can be kept set to UTC (Universal Coordinated Time), the time system that is the same everywhere in the world.

QSL cards for exchange with other hams in acknowledgment of a contact are not really an accessory. But a personal QSL card is something that every ham wants to have. The card can be very simple or very elaborate, depending on personal tastes and budget. A number of printers specializing in QSL cards advertise in amateur radio magazines.

To keep the operating area of the station clear for using the key and for logging and copying messages, many hams have found that it is convenient to locate accessory gear on a shelf over the transmitter and receiver. There they are easy to reach and out of the way.

Before putting the new station on the air, two things must be done: the transmitter must be "neutralized" and tuned, two procedures quite different from turning on a broadcast radio receiver. The electrical characteristics of the amplifier tubes in a transmitter cause some of the output to be fed back into the input, which can produce a

FIG. 6-18 The 24-hour Universal Coordinated Time system is used by amateurs around the world. This clock reads 1511 UTC.

spurious signal on an unwanted frequency. Neutralization prevents this. There are several amplifiers in a transmitter, and each may have to be neutralized; neutralization is a "set-and-forget" adjustment that usually must be repeated only when tubes in any of the neutralized stages are replaced. The procedure varies depending on the particular transmitter. If an instruction manual is not available, an experienced ham can demonstrate how neutralization is accomplished.

Under FCC regulations the station licensee is responsible for a clean signal that will not cause interference. The signal must be stable in frequency, pure in tone, and free of harmonics, parasitics, and other spurious signals. That is one reason that neutralization and proper tuning are important; common courtesy is another. At first, transmitter tuning seems like a very complicated process, but it soon becomes second nature. Here again, an Elmer can be very helpful in overcoming mysteries.

Tuning a Transmitter

Whenever the transmitting band is changed, several controls must be adjusted to pass the newly selected frequency through each of the tuned circuits. Which controls and how many must be adjusted depend on the brand and model of the transmitter. Tuning should be done only with the transmitter connected into a dummy load, never over the air, where it could cause interference. Generally, the procedure is as follows:

- Set the *bandswitch* to the desired band—10, 15, 40, or 80 meters for a Novice.
- Set the *tuning dial* to the desired frequency.
- Turn the *drive* or *microphone level* control fully counterclockwise.
- Set the *meter function switch* to *plate current.*
- Turn the *mode switch* to the *tune* position.
- Slowly turn the *drive* or *microphone level* control clockwise until there is a slight indication of plate current.
- Adjust the *preselector* or *grid tuning* control for maximum plate current.
- Adjust the *plate tuning* control for minimum plate current. There should be a pronounced *dip* in the plate current reading as the control knob is turned.
- Increase the drive level by turning the *drive* or *microphone level* control clockwise.
- Readjust the *plate tuning* control for minimum plate current.
- Increase the *drive* again until further clockwise rotation produces no increase in plate current. Then readjust the *plate tuning* control for minimum plate current.
- Adjust the *antenna loading* control for maximum plate current.
- Readjust the *plate tuning* for minimum plate current.
- Readjust the *antenna loading* control for maximum plate current.
- Repeat the *plate tuning* and *antenna loading* control adjustments until there is no further change in plate current.
- Set the *mode switch* to *standby.*

- Disconnect the transmitter from the dummy load and connect it to the antenna. Set the *mode switch* to CW.

Always keep key-down time during tuneup very brief; do not hold the key down more than 10 seconds at a time to avoid possible damage to the final amplifier.

Choosing a band on which to operate is obviously limited by the station antenna and its capabilities. To select a frequency within the band, the proper procedure is first to listen across the authorized Novice segment of the band either for a CQ (a call to answer) or a clear frequency on which to transmit a CQ. FCC regulations require station identification at the beginning of each transmission; a typical QSO or amateur contact is described in Chapter 5.

Listen First

If standard amateur radio operating practices are unfamiliar, it is a good idea to listen to some on-the-air conversations before initiating a first contact. No one expects a flawless performance from a Novice; every ham has made his or her own share of mistakes at first. One of the reasons specific frequency segments have been designated for Novice use is to encourage newcomers to operate without worry or embarrassment. So, like learning to swim, the best way to get started in amateur radio is often to jump right in.

A widely distributed ARRL pamphlet, *Operating an Amateur Radio Station*, has this to say about establishing contact:

> The best way to do this, especially at first, is to listen until you hear someone calling CQ, and call them. This requires a little patience, but that's something else all amateurs must learn if we are to share our bands in harmony. Tune around near your own frequency. If you hear a CQ, put your transmitter on that frequency (without putting a signal on the air), wait until the operator indicates he or she is listening, then call thus: W6ZRJ W6ZRJ DE W7PGY W7PGY $\overline{\text{AR}}$. If no answer (to anyone), this may be repeated; brief repeated calls are preferred to long drawn out ones. Chances are, if they are to hear you at all, they will hear your first brief call; most amateurs seldom tune far

from their transmitting frequency to listen after a CQ. Note the ending signals (Chapter 5). These have a special significance of their own to indicate to a casual listener the "status of the contact."

In answer to your call (assuming you are heard), the called station will reply: W7PGY DE W6ZRJ R . . . and then go into conversation. That R (for ROGER) means that your call was received correctly. That's all it means—RECEIVED. It does not mean correct, I agree or I will comply. It is not sent unless everything was received correctly.

There are no legal limits to what amateurs may talk about on the air other than prohibitions concerning profanity, obscenity, and

FIG. 6-19 After meeting on the air, these two hams have gotten together several times to extend their friendship. Dr. Mark Woodward, WB9NNO, is a veterinarian from Cambridge City, Indiana; his amateur friend, on the right, lives in Bremen, West Germany. When the American ham visited Ulf-Dietmar Ernst, DK9KR, the German showed him the Rhine Valley from the back seat of a motorcycle.

conversation relating to business affairs. Most amateurs try to avoid controversial subjects and observe good taste, remembering that the airwaves are open to anybody and others are probably listening.

A New Zealand amateur says: "I was told when I was a kid fiddling around with crystal sets that if I took on radio for a hobby I would never be short of friends, and that's just what it has all added up to." He cites an on-the-air experience while he was trying to call a colleague in Bolivia, using his elementary Spanish. "It was an abortive call, but a Russian sitting in his radio shack somewhere in the Ukraine recognized my ZL3JO call sign even as it was spelled out in Spanish, and came homing in for a 5-minute chat."

That's amateur radio!

7

Many Roads to Follow

"To challenge the limits of human endurance." That is how Japanese radio amateur Naomi Uemura, JG1QFW, expressed his reason for attempting to reach the North Pole alone. He was to become the first person ever to accomplish the feat. During his 55-day solo trek across the Arctic wilderness by dog sled, climaxed by his arrival at the Pole on April 29, 1978, Uemura relied on daily ham radio contact with Japanese, North American, and other amateur stations.

When the organizers of the Olympic Torch Run that was to take the symbolic flame from Virginia to Lake Placid in 1980 saw the need for communications help, it didn't take them long to think of amateur radio. Scores of amateurs volunteered to relay progress reports and word of poor weather conditions along the winding route. In part because of the well-prepared communications network, the torch made it to the site of the XIII Winter Games without a hitch.

Such examples show that amateur radio has been used to maintain communications during highly unusual and record-setting feats. But those who monitor the amateur bands may suddenly and unexpectedly find themselves smack in the middle of an exotic event. Their unique communication capability may even allow hams to play an active role in making history. In April 1979, 13-year-old Mike Davis, WD6FFV, of Torrance, California, overheard a distress call from the *Carmen*, a fishing boat that was sinking near Jamaica. Reacting quickly despite the late hour (1:00 A.M. in California), Mike notified the Coast Guard in Long Beach, which in turn contacted their base in Miami. Mike kept in contact with the sinking ship while the search continued. The *Carmen* was finally towed to safety by a rescue craft, thanks to the alertness of a ham on the other side of the continent.

On a less-glamorous scale, hams routinely provide the communications link for parades, sports-car rallies, marathons, boat races, and other community events. They also use their radio hobby as an adjunct to other hobbies and special interests (but never in conjunction with business or commercial activities—that's prohibited). Stamp collectors, chess players, bridge groups, hikers and campers, mountain climbers, model railroad addicts, model builders, and others have all found that amateur radio adds both fun and utility to such activities.

Amateur radio itself has become so diversified that many hams find themselves deeply if not exclusively involved in some form of specialization. A sampling of the amateur bands during any popular operating hours may reveal what amounts to a smorgasbord. With a flick of the switch, a turn of the dial, listeners may find themselves in the midst of a DX contact, the frenzy of a contest, a message being relayed, a summons for help after a road accident, an OSCAR satellite pass, a technical discussion, or just a casual "rag chew."

What will become a newcomer's specialty? The collector may opt for DXing—collecting contacts with faraway places and the QSL cards that are a record of those contacts. For the competitive it may be contests—matching skills and determination with other hams near and far. For the gadgeteer it could be checking out new accessories at ham conventions and in the pages of amateur radio journals. For those who like to work with their hands, there is new gear to be

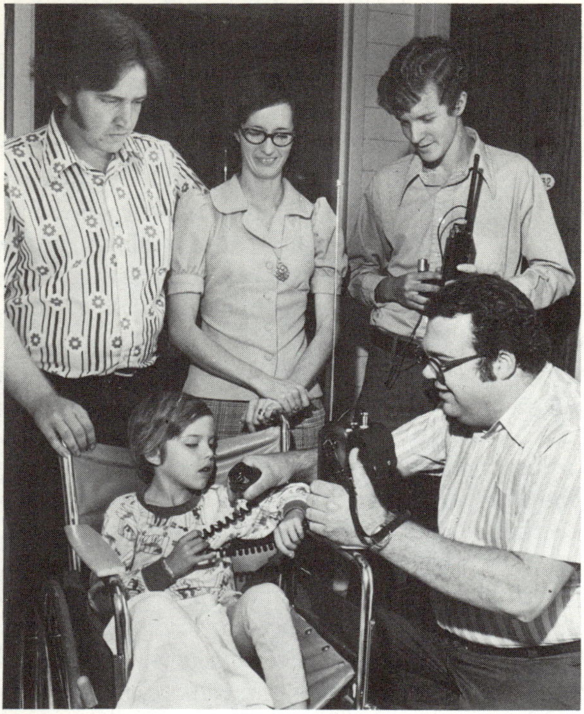

FIG. 7-1 Each holiday season, many hospitalized children pass their important hints to Santa with help from volunteer amateur radio operators.

built and kits to be assembled. For the adventurous, there are DXpeditions and Field Day. For those with a desire to serve, there is an extensive public service organization. For those who take pride in skills, there are awards for excellence and perseverance. For those who like to experiment, there are the OSCAR satellites, moonbounce, slow-scan TV, and the ultra-high-frequency amateur bands. And for someone who just likes to make new friends, there is the never-ending feeling of anticipation and challenge each time a CQ is sent out over the airwaves. Who will answer this time? An old friend or a new one? Someone from across town or on the other side of the world?

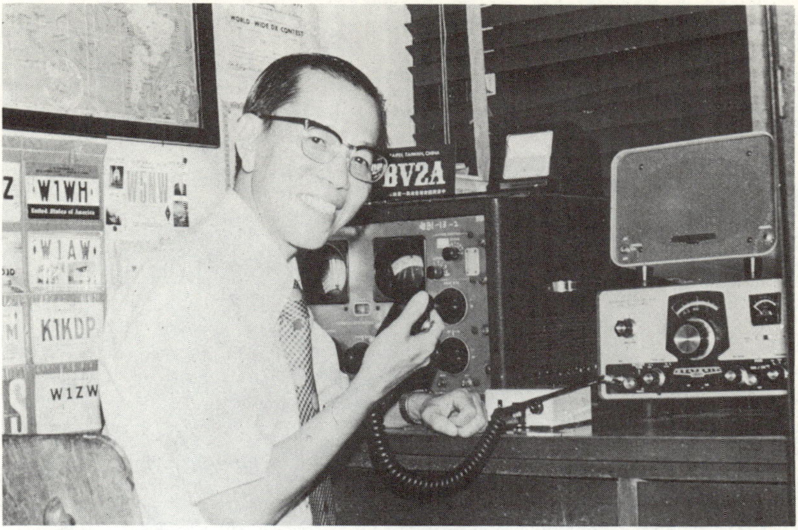

FIG. 7-2 When he speaks, DXers listen. Tim Chen, BV2A, offers DX operators a rare radio glimpse of Taiwan.

Contesting: A Walk on the Wild Side

Say "contest" to almost any ham and the first response is apt to be "Field Day." It is the oldest, probably the biggest, and certainly the most famous of all amateur radio operating activities. To call Field Day a contest is almost like calling the Taj Mahal a building, for it is much, much more. The following recollections provide a taste of what Field Day is all about.

> All in all, it was an excellent Field Day on Bald Mountain Summit, just east of Mt. Rainier. The weather was bad; a 40-mph wind blew continuously and we had some light snow. Despite the weather, all the antennas stayed up, the generators kept generating, and the coffee was very popular. When the weather broke on Sunday we had breathtaking views in all directions.
>
> Last year we went the high-power route and had no end of troubles. This year we left the amplifiers at home, entered the

FIG. 7-3 Perhaps the most popular of all ham radio activities, Field Day is a combination of contest, weekend in the country, and ham ingenuity. This youngster has found an ideal spot for Field Day—the middle of a field!

> low-power category; that cut our troubles in half and doubled our total score.
>
> Another great Field Day. More attended this year to take our annual ham radio voyage on Lake Michigan. And, of course, Murphy [a traditional term for unexpected difficulties] didn't miss the boat. He got us first by letting Bruce's (K9EEA) 46-foot yacht run aground just north of Chicago. Our operator on duty at the time kept at the key throughout, except to poke his head out now and then and ask nervously, "Are we abandoning ship?"
>
> With little funds and less brains, we got the great idea of using a lawn-mower engine, an automobile generator, and a retired battery for our power plant. We mounted all these components on a plank and hooked up the wiring. Not a single microamp. So Bill kicked it at just the right spot, polarizing the fields, and it worked.

If these accounts make Field Day sound like a zany, undisciplined escape from reality, the conclusion is understandable. But

Field Day really is more. Its original purpose, as envisioned by ARRL, was to provide a large-scale exercise in preparedness for emergency communications, and that element is still very much alive. Groups of hams, usually local radio clubs, set up portable stations at sites ranging from remote mountaintops to the middle of a city park, and attempt, operating continuously over a June weekend, to make as many contacts as possible with similar groups operating across the United States and Canada. It is a combination of serious effort, get-away-from-it-all camp-out, two-day party, and sheer good fellowship.

Field Day rules are as simple as possible. For the most part, only noncommercial power is used, which means that participating stations are operated with power from either a portable generator or batteries. There are different classes of entrants—club groups that set up at a temporary site specifically for Field Day (the largest class), nonclub stations set up and operated by not more than two licensed amateurs, stations located in and operated from a vehicle, and fixed stations using portable power. Extra credit is awarded for Novice contacts, for CW (Morse code) contacts, for contacts using low power (200 watts or less) or natural (non-petroleum-based) power, and for contacts through an OSCAR satellite.

The emergency-preparedness aspect of Field Day may take on a note of reality. A forest fire erupted in the Jemez Mountains in New Mexico just as the Los Alamos Amateur Radio Club was preparing to set up its communications van and generators. Their Field Day site was Graduation Flats—and Graduation Flats just happened to be selected by the Forest Service for its base fire camp. The radio van and equipment were quickly pressed into service to supply power and communications support. Over an 8-day period, a number of club members put in long hours, primarily in providing a direct link with the outside telephone system through several mountaintop amateur repeaters. Since the Forest Service equipment did not have telephone-access capability, the amateur's round-the-clock operation provided coordination for food supplies, firefighting equipment, supplies, weather information, and maintenance sources, as well as communications with families and relatives of people in the stricken area. When Field Day finally began, the club was more than ready to swing into action!

Contesting for Novices is limited, of course, by the bands they

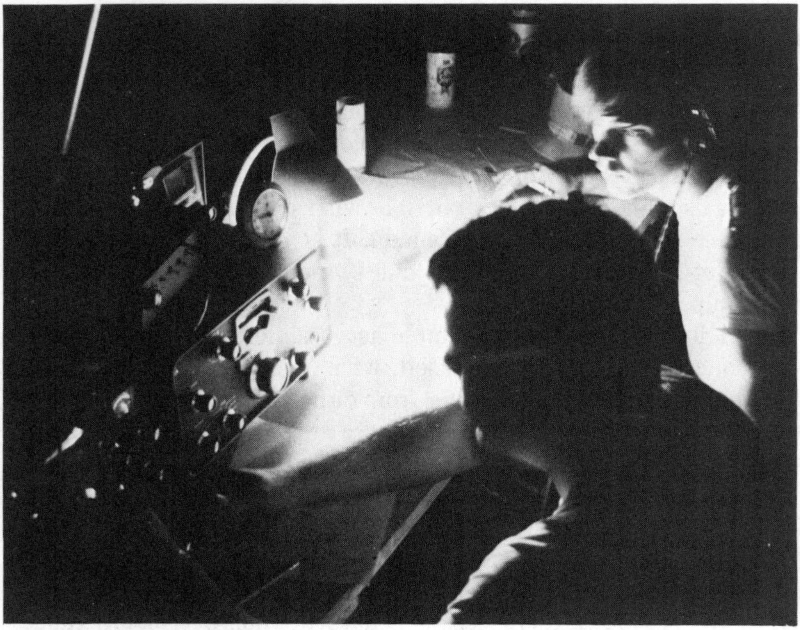

FIG. 7-4 To garner the maximum number of points, Field Day operators often keep their stations running all night.

are authorized to use. But there are a number of opportunities for very active Novice involvement—such as Field Day—plus a contest specifically for Novices called the Novice Roundup. Sponsored by ARRL, it occurs annually for 10 days in February. The objective is to make as many contacts as possible in a maximum of 30 operating hours. Participation in the NR is a highlight of many Novice careers.

For the amateur looking for something really exotic and challenging in a contest, there is even a "moonbounce" contest. It is ARRL's International EME (earth–moon–earth) Competition. To qualify, all contacts must be made via an earth–moon–earth path on a frequency above 50 MHz. What is "earth–moon–earth"? Operators point their antennas up—and bounce their signals off the moon. This is a contest for skilled operators with very good antenna systems!

The method of scoring varies from one contest to another, but most are based on a "multiplier" system. The total number of contacts made is multiplied by the number of countries or number of states worked. A winning station in an international contest may roll up more than 1 million points. Eligibility requirements, rules, and scoring systems for all the contests are published by the sponsoring organizations in the amateur journals. For the serious competitor, contesting is virtually a full-time amateur activity year round. Among the top contest attractions are these:

January	ARRL VHF Sweepstakes
	CQ 160-Meter Contest (CW)
February	ARRL Novice Roundup
	ARRL International DX Competition (CW)
March	ARRL International DX Competition (phone)
	CQ WPX Contest (phone)
May	CQ WPX Contest (CW)
June	ARRL Field Day
	ARRL VHF QSO Party
	All-Asian Phone Contest
July	International Amateur Radio Union Radiosport Championship
August	Worked-All-Europe CW Contest
	All-Asian CW Contest
September	ARRL VHF QSO Party
	Worked-All-Europe Phone Contest
October	CQ Worldwide DX Phone Contest
November	ARRL Sweepstakes
	CQ Worldwide DX CW Contest
December	ARRL 160-Meter Contest
	ARRL 10-Meter Contest

The Lure of DX

Kingman Reef is a tiny coral atoll in the South Pacific Ocean. Four hams—members of the Northern California DX Club—spent days searching for the uninhabited little island aboard the motor-sailer *Caroline*. Once there, they endured a gale-force tropical storm—all to set up and operate at a "rare" location for just 29 hours. In that period they logged 5535 contacts with avid DXers around the world anxious to add another "country" to the roster of those they've contacted. It was another successful "DXpedition," an activity in which groups of hams—whose sanity might be questioned by non-amateurs!—make their way to out-of-the-way places where there is no ham radio.

DXpeditions are perhaps the ultimate in DX; "ordinary" DX is simpler. There are, for amateur radio purposes, some 300 or more places considered to be "countries." When a ham has verified contact with 100 or more of these, he is eligible for what is probably amateur radio's most prestigious award, a DX Century Club (DXCC) certificate from ARRL. Endorsements to the original certificate are awarded for further achievement levels. Top DXers are listed regularly in the DX Honor Roll in *QST*, ARRL's monthly journal.

FIG. 7-5 The first 100 countries worked and confirmed via QSL cards, necessary to qualify for the DX Century Club certificate, are usually just the beginning of a DXer's career.

When a DXer has achieved whatever his or her goal may be in DXCC, there are still other worlds to conquer—such as Five-Band DXCC (100 verified "countries" on each of five amateur bands), the Satellite Achievement Certificate (contacts through an OSCAR satellite), and DXCC on the 160-meter band or using radioteletype.

In a sense, DXers are like stamp or coin collectors—always anxious to acquire a contact they have not had before, especially a rare one. The tangible form of a DX "collection" is an array of QSL card verifications from exotic countries all over the world—first on the walls of the ham shack and, as that space is exhausted, in file containers. Unlike most stamp or coin collectors, however, DXers enjoy another dimension to their hobby—the international friendships DXing produces. Some of these can and do grow into life-long relationships through regular on-the-air schedules. One New England ham maintained weekly schedules with several friends in Australia for more than 30 years; to his wife's frequent dismay, it often interfered with her weekend planning. After his retirement, she somewhat reluctantly agreed to travel to Australia to meet these unseen friends. To her astonishment, they were greeted with a reception and round of entertainment that would have done justice to a head of state. In that particular world of ham radio, her husband was a celebrity. The story is not unique; there is no such thing as "foreign" travel for a radio amateur who has worked DX even to a modest extent.

DX exemplifies all that is romantic and glamorous in amateur radio. It appeals to the armchair traveler, the geographer, and the internationalist. No other human activity regularly involves so many people in so many of the earth's places in direct, person-to-person communication. DXing can be thought of as an instrument for better international understanding and world friendship.

What, actually, is DX? To a new ham it may be the next state; experienced amateurs may be looking only for "rare" countries. That first contact with a station in another part of the country or the world is a very exciting experience indeed. DX operating—contacting stations at distant locations—is accomplished primarily on the high frequencies, the bands 80 through 10 meters. The VHF and UHF bands are usually "line-of-sight"; that is, their signals usually travel within the optical horizon. Intercontinental contact is, therefore, generally impossible except during periods of high sunspot activity or through an OSCAR satellite. DX may also be considered dependent on power and antenna configuration. A QSO with a station in a distant country via a long wire antenna and 5 watts of power is an impressive DX accomplishment.

Twenty meters is traditionally the DXers' favorite band. International signals are always coming in from somewhere, and under ideal conditions DX is workable around the clock on CW, phone, radioteletype, and slow-scan TV. Fifteen meters is almost always open to South America, Europe, and Africa from the East Coast of North America, and to the Pacific and Southeast Asia from the West Coast. Forty meters can be a good nighttime DX band. Ten meters is an excellent daytime DX band when sunspot activity is high, but even during times of inactivity, exciting openings can occur. Most serious DXers use beam antennas that can be rotated to point them in the right direction for particular band conditions, but the newcomer will find that he or she can work some DX very satisfactorily with a simple dipole antenna.

Awards

Hams call them "wallpaper," but display them proudly as diplomas. Amateur radio is one pursuit where both proficiency and accomplishment are visibly recognized by a number of different award certificates. The typical ham shack display may begin with a framed ARRL membership certificate and progress through RCC, WAS, and WAC to DXCC.

RCC is the Rag Chewer's Club certificate, awarded by ARRL. To achieve it, an amateur need only maintain a contact for a solid half hour, and report the QSO to ARRL.

WAS is Worked All States, probably the most popular of all operating achievement awards. It is issued by ARRL on receipt of verification (QSL cards) of contact with an amateur station in each of the 50 states. In a typical year, about 10% of all WAS awards are made to Novices.

WAC is Worked All Continents, sponsored by the Interna-

FIG. 7-6 Just 2 years on the air netted WA2GOQ his Worked All States award at age 16.

tional Amateur Radio Union through ARRL. It requires verified contact with an amateur station on each of the world's six continents.

ARRL's headquarters station, W1AW, awards code-proficiency certificates for copying Morse code on their Qualifying Runs, transmitted at speeds from 10 through 40 words per minute.

ARRL's A-1 Operator Club is unusual in that it cannot be applied for. Club members submit nominations for those they consider top-notch amateur radio operators.

Other amateur radio organizations around the world sponsor awards of their own, many of which are available to American hams who qualify. Amateur radio publications regularly report details on new and existing awards.

Message Handling

One of the very earliest forms of amateur radio public service was the transmission of messages on behalf of individuals and charitable organizations. Still a popular pursuit among a large body of hams, it has been described as "fun with a purpose."

One form of message handling is performed by hams individually—putting service men and women around the world in touch with their families. This is done by means of a "phone patch" that connects the U.S. amateur's station to the telephone system, enabling service personnel overseas, using amateur stations at a base or aboard ship, to talk with their families via telephone. During the Vietnam conflict, hundreds of U.S. hams spent countless hours, day after day, providing phone patches for thousands of soldiers, sailors, and airmen. Senator Barry Goldwater and a group of other local amateurs ran phone patches through his station in Phoenix, Arizona, around the clock for the duration of the conflict.

The interest shared by the amateurs specializing in written message handling, however, is simply called "traffic." They operate on a regular basis through one or more of a series of networks or "nets" from coast to coast. These nets are part of an ARRL-sponsored organization known as the National Traffic System (NTS). Local nets are interlocked with section, state, and region nets, which, in turn, mesh with an area net. The area nets are linked by an elite group of traffic operators known as the Transcontinental Corps. A message originated locally is passed up the line in a standard procedure and down again to a station near its destination. A message that is local in both origin and destination is, of course, handled wholly within the local net. Net members meet on the air on a regular schedule; the passing of messages is coordinated by one member at each level serving as a net control station. A regular net is very much like a club—it meets at a regular time at a certain place (frequency) and is presided over by a master of ceremonies (net control) to conduct a particular kind of activity (the passing of traffic).

There are many kinds of nets. Informal nets are composed of groups of hams who just like to get together on the air for a roundtable. There are special interest nets of stamp collectors, antique car aficionados, mountain climbers, and so on; there are nets sponsored by technical groups interested in such areas as antennas, low-power operation, and DXing. An amateur can check into phone nets, CW nets, slow-scan television nets, or radioteletype nets!

A traffic net is something else. Its purpose is to provide the public with a service while its members develop their operating skills and have a good time in the process. It functions as part of the National Traffic System using formalized procedures developed over the years for speed, accuracy, and simplicity.

Who can send a message by amateur radio? Just about anyone. A message may concern any subject as long as it is not highly controversial or related to a business or commercial interest. And sending one is not quite like calling Western Union, to be sure. First, a nonham must locate an amateur with access to a traffic net—which should not be difficult. If an amateur is not familiar with a local net, the chances are that he or she knows someone who is. There is, of course, no charge for sending a message by amateur radio. Because delivery cannot be guaranteed, other means of public communications should be used to get a critically important message through, but the amateur traffic performance record is very good indeed. The National Traffic System operates on a regular, scheduled basis, so almost all messages are delivered within a reasonable time. The message will be received by an amateur in a local net within toll-free telephone calling distance of the addressee. The message will be delivered by telephone if possible or, if not, on an official ARRL

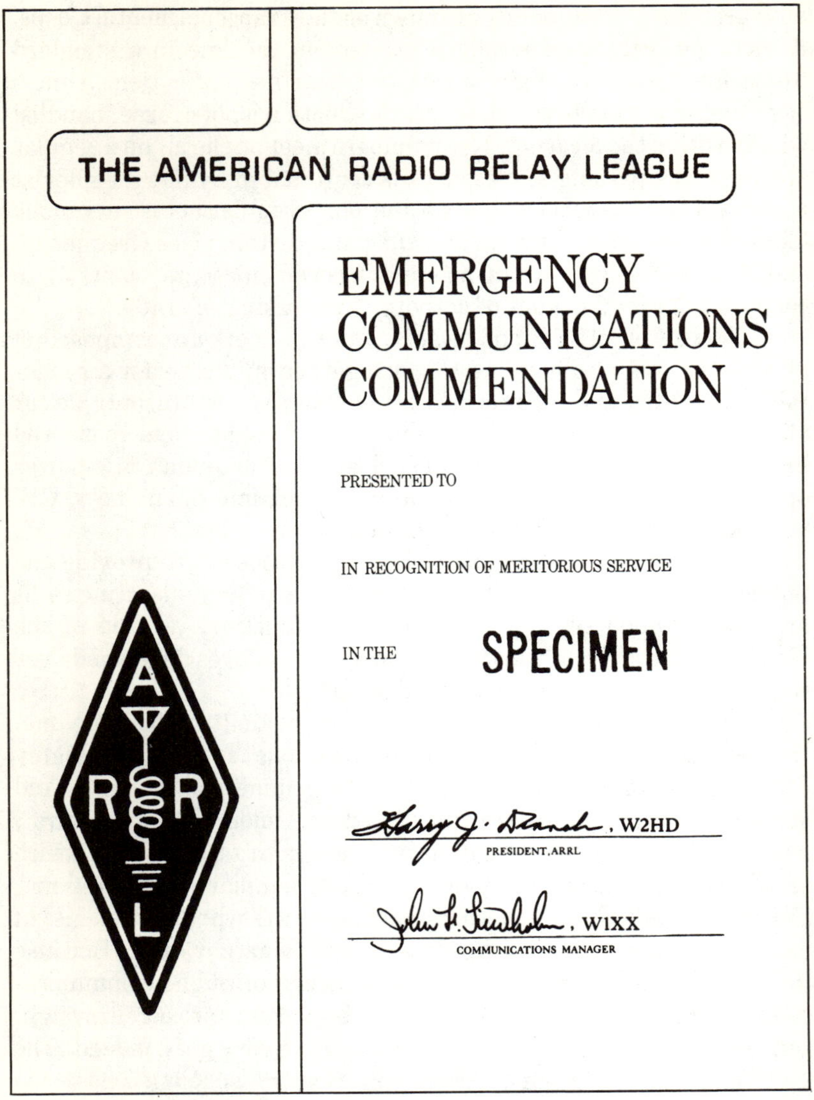
THE AMERICAN RADIO RELAY LEAGUE

EMERGENCY COMMUNICATIONS COMMENDATION

PRESENTED TO

IN RECOGNITION OF MERITORIOUS SERVICE

IN THE SPECIMEN

, W2HD
PRESIDENT, ARRL

, W1XX
COMMUNICATIONS MANAGER

FIG. 7-7 Amateurs who perform voluntary communications during an emergency are eligible for this handsome award from ARRL.

Radiogram form that looks very much like a telegram. The delivering amateur will also usually be glad to handle a reply or acknowledgment.

The amateur National Traffic System functions day in and day out, 365 days a year. In an emergency situation, such as the disruption of commercial communications facilities by a hurricane, earthquake, or flood, net operations are stepped up and the number of participants swells. Messages are classified according to a strict priority system, and hams will go to great lengths to assure that those of an emergency nature are delivered expeditiously.

The National Traffic System has many component parts in the form of local, region, section, and area nets. But NTS itself is only one part of the ARRL's field organization. Another is the Amateur Radio Emergency Service, ARES. When disaster strikes, close coordination between NTS and ARES sets the whole machine running in high gear to provide communications when and where needed. Volunteers are welcome at every level of this organizational structure, from Section Communications Manager to Net Manager, from Section Emergency Coordinator through local Emergency Coordinators—even individual station appointments such as Official Traffic Station and Official Emergency Station. For the new amateur interested in becoming a part of the NTS or ARES, the Section Communications Manager is a good person to contact. They are listed in *QST*, the ARRL monthly journal. Otherwise, an inquiry to the ARRL Communications Department, Newington, Connecticut 06111, will elicit the names of local officials.

Repeaters—the relay devices operated by local amateur groups to extend the range of 2-meter FM communications—are widely used to provide service to the public. On a purely local level, repeater groups may supply weather information to the Weather Service and broadcast stations. They provide auxiliary communications for the police and coordination of many charitable events, such as Walk-athons and Bike-athons. The small, hand-held transceivers many hams use for 2-meter FM make such communications very flexible and efficient.

More detailed information about message handling and traffic operations appears in three ARRL publications, *Operating an*

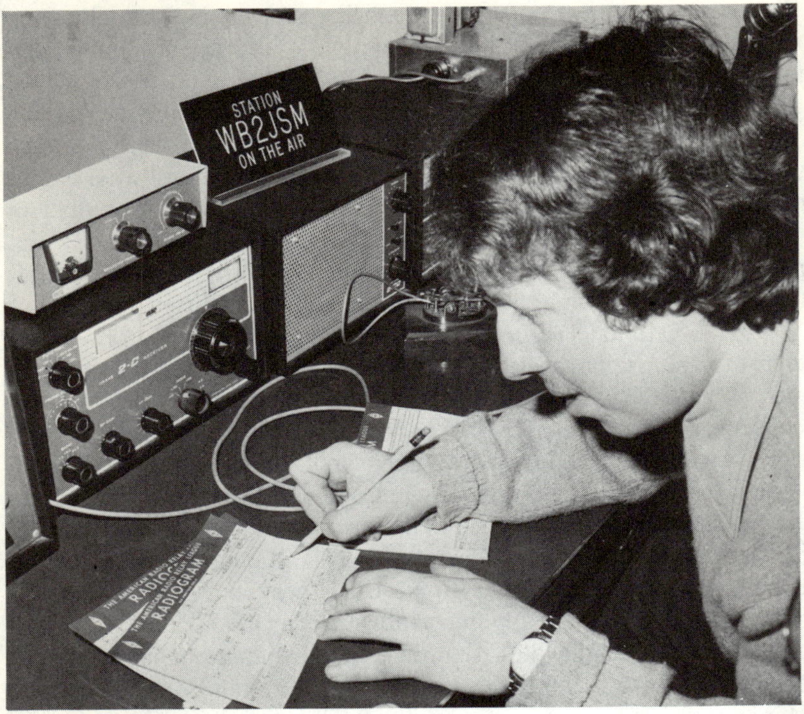

FIG. 7-8 A message received directly from Guatemala, where a disastrous earthquake has occurred, is checked for accuracy at WB2JSM, the amateur radio station at a museum in New York City. Radio amateurs train to provide emergency communications in the event of a natural disaster, power failure, or other disruption of vital communications.

Amateur Radio Station, the *Public Service Communications Manual*, and the *ARRL Net Directory*. All are available, free of charge, from ARRL.

VHF/UHF—New Horizons

Historically, amateur radio has been identified primarily with the high frequencies—the 80- through 10-meter bands. Use of the VHF/UHF bands has been a relatively recent and continuing development. These are the bands in the frequencies above 50 MHz, from 50 through 2450 MHz. They include 2-meter FM and had always been thought of as line-of-sight bands. But that is changing. Developments such as the OSCAR satellites are making it possible to use VHF/UHF over some surprisingly long distances. Amateur experimenters are busily engaged in extending the performance capabilities of upper-frequency transmitters, receivers, and antenna systems. They have discovered natural phenomena which, under the right conditions, enhance propagation of VHF/UHF signals to provide nationwide and even worldwide coverage. These developments and opportunities, still largely unexplored, make the world above 50 MHz especially exciting and challenging for amateurs seeking new frontiers. At VHF/UHF, both technical and operating skills are challenged to the utmost.

Trying to reach ever-more-distant stations has been a key part of amateur radio's appeal from the beginning. Working DX on the VHF/UHF bands, however, is more a challenge of person against nature than a person-against-person struggle, as on the HF bands. There is an especially close feeling of cooperation and fraternalism

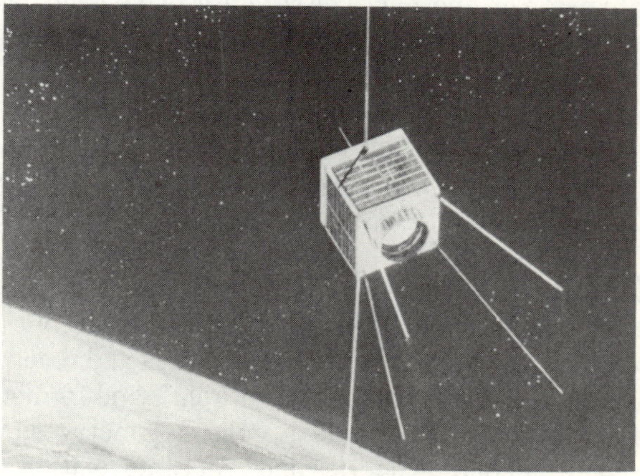

FIG. 7-9 The eighth in the series of amateur radio satellites, AMSAT-OSCAR 8, has circled the earth, providing reliable point-to-point communications, since 1978. Mike Smithwick, W6TUF, rendered this view.

among the VHF/UHFers in exchanging knowledge and pooling resources. It is one specialty where operator-built or "homebrew" gear, especially antennas, is still very prevalent.

Regular amateur radio magazine columns and special manuals deal with VHF/UHF. VHF/UHFers even have their own nets and contests. A knowledge of propagation, of sporadic-E effects, ionospheric and meteor scatter, and other special phenomena are basic tools of the trade.

VHF/UHF is a world that Novice licensees may only look forward to, since they are limited to the 10-, 15-, 40-, and 80-meter HF bands. But Novices need take just one step up the licensing ladder, to a Technician license, to gain access to all the VHF/UHF bands.

OSCAR—Amateur Radio in the Space Age

An OSCAR satellite is an *O*rbiting *S*atellite *C*arrying *A*mateur *R*adio. It carries one or more transponders—each a special kind of repeater—receiving signals on one frequency and retransmitting them back to earth on another. The satellites have been designed so they can be used by most amateurs with ordinary or readily available equipment.

What makes satellite communication special for a growing body of hams is the opportunity for firsthand participation in the space age. A contact through a tiny space vehicle orbiting out there with some highly sophisticated gadgetry has an appeal all its own.

OSCAR satellites have been used for some rather exotic experiments, but most of the OSCAR activity by thousands of amateurs in more than 100 countries has been with CW and single-sideband phone, most frequently transmitting on 2 meters and receiving on 10 meters. (These are referred to as 2-meter uplink and 10-meter downlink.) The satellite also transmits regular telemetry signals that report the condition of its batteries and other information. Access is controlled by a number of satellite command stations whose operators can turn OSCAR on and off according to maintenance requirements.

The only requirements for hearing OSCAR concern knowing

when and where to tune in. The orbital paths are predictable; schedules are published regularly in the amateur publications. From these, the time and point of access for any location can be calculated easily. There is even a device called an OSCARLOCATOR, available from ARRL, which makes pinpointing an OSCAR even simpler. W1AW, ARRL's headquarters station, also transmits daily OSCAR orbital bulletins. Although OSCAR can be heard with a simple dipole antenna, most serious users rely on beam antennas that can be pointed to the expected OSCAR path.

A Technician or higher-class license is required to operate on OSCAR frequencies, but a Novice can easily listen to the satellite—and listening is an excellent way to get started in satellite communi-

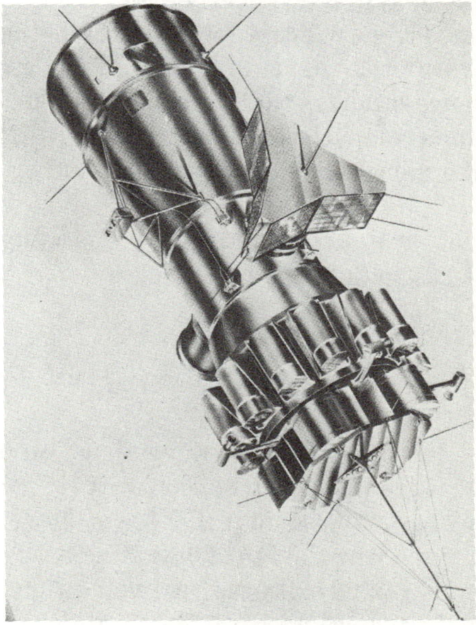

FIG. 7-10 An artist's conception of AMSAT-Phase III in space, before separation from the launch vehicle. This satellite, the ninth in the series that have been designed and built by and for amateur radio operators, will expand the horizons of those who use them. Contacts up to 11,000 miles are possible through this satellite. (European Space Agency photo.)

cation. Tuning in is a good way to become familiar with some of the characteristics peculiar to OSCAR signals. Signal strength is affected by the distance of the satellite, which changes, of course, as it approaches and recedes.

Another OSCAR milestone will soon be reached with the launch of AMSAT-Phase III. Unlike its low-orbiting cousins, this satellite will trace an elliptical orbit around the earth. Its apogee (highest point) will be about 23,600 miles above us. This means that amateurs using the satellite can talk through its transponder for much longer periods of time (up to 10½ hours). In addition, users won't have to worry about aiming their antennas every few minutes. Perhaps the most exciting aspect of the latest OSCAR is that amateurs will be able to reach nearly every point on the face of the earth during at least one of every few orbits. Whether OSCAR users enjoy talking with other hams in distant lands, assisting students with space science studies, passing emergency messages, or even linking their home computers across continents, AMSAT-Phase III will offer them a more reliable, convenient, and *predictable* way of doing it.

Two Soviet amateur satellites have operating characteristics similar in most respects to the OSCARs.

QRP—The Challenge of Low Power

Amateur radio operators—depending on their class of license—are permitted to use up to 1000 watts of power, though few actually do. Novices may use up to 250 watts. More and more hams, however, are discovering the fun of operating with only 10 or even 5 watts. It takes skill and patience, as well as ingenuity, to work through all the "big guns" with a little "flea power" outfit. But the QRPers, as they are known (after the Q-signal for "reduced power"), are accomplishing very impressive results, including some surprising DX, with such rigs. In addition, such equipment has the advantages of portability, simplicity, and low cost. An entire QRP station—rig, key, and antenna—will fit into a suitcase for vacation travel, or into a backpack for a climb up a mountain!

FIG. 7-11 The challenges of low-power (QRP) amateur radio and mountaineering were combined when two California hams took to the airwaves from the summit of 14,496-foot Mount Whitney. (Photo by WA6VBA.)

With a 7-watt station set up on a West Indies island, one ardent QRPer made contacts coast to coast in the United States and Canada, and then went on to work Europe, South America, and even Japan.

It does, to be sure, take a certain amount of skill to sneak a relatively weak signal around and between higher-powered stations, but it is also very satisfying to have an English ham report, "Your signal is really booming in over here" when one has been running just 3 watts into a beam antenna!

Join a Club

Just deciding which amateur specialty to pursue is challenging enough for the newcomer, let alone learning all the special tricks of the trade each specialty demands. Membership in a local amateur radio club affords friendships and expertise invaluable to the new ham as he or she meets those challenges.

Some amateur radio clubs are specialty organizations

themselves—organized by and for DXers, contesters, or SSTV enthusiasts. But most local clubs are general-interest groups, with members interested in most or all of the various amateur activities, including nonoperating interests such as amateur politics! In such a club, the newcomer will be quickly exposed to all the magic and appeal of many possible interests. New hams will have opportunities for firsthand experience in many amateur radio specialties and to get expert guidance in those they elect to pursue.

Similarly, membership in the American Radio Relay League provides access to information about all aspects of amateur radio through *QST*, the League's monthly journal, its library of publications on just about every specialty, and its Technical Information Service. New Novice members will make contact with other

FIG. 7-12 Prospective Novices take their first big step toward earning their first amateur license in classes sponsored by the local clubs and the American Radio Relay League. (NASA photo.)

FIG. 7-13 Want to help master the Morse code? Want to hear up-to-the-minute bulletins of interest to all amateurs? Then you'll want to tune in W1AW, ARRL's Maxim Memorial Station in Newington, Connecticut. Schedules are available from ARRL headquarters.

amateurs everywhere who share their particular interests. They will keep on top of new developments in all amateur fields. And through their membership in the League, they will be supporting amateur radio's strongest collective voice in regulatory decisions at home and abroad.

8

Getting Into the Act

Hams don't often just sit back and "let the other guy do it"—they are willing to lend a hand whenever and wherever their help may be needed. One newly licensed amateur wrote the following account of his first experience with amateur radio public service:

> One evening I was tuning the 80-meter band looking for someone sending code slowly enough for me to copy. Ah. A good, loud station, just the right speed, and he was finishing his contact. Full of confidence, I immediately called him and he came right back with a reply. Great. But then he began asking me strange questions like QRU? I didn't even know what he meant. So I sent a question mark. He came back slowly and patiently, asking if I had any traffic. I said no. I got a polite thanks, a welcome to the net, and was told to stand by. Good grief, I had broken in on a Net Control Station and had checked into a traffic net!

Adverse weather conditions seem to spawn amateurs who use lightweight, portable transceivers to assist local authorities during and after serious storms. The Birmingham Amateur Radio Emergency Service (BARES) in Alabama works very closely with Civil Defense and Red Cross officials; its club station, in fact, is located in the Red Cross headquarters building. On a January morning, a tornado struck McComb and heavy thunderstorms began crossing the Alabama–Mississippi border. The BARES station and the stations at Jefferson County Civil Defense Emergency Operation Center were activated, and emergency nets on 2 and 75 meters were placed on standby status. As reports of the bad weather came in, the information was plotted on maps at the headquarters. Tornado touchdowns were reported in sparsely populated areas in western Alabama. Suddenly, the call "Break Emergency!" crackled over the air.

"This is K4TQR mobile four. I'm in the middle of a tornado touchdown right now!"

Then W4TWK reported, "A tornado just passed over my house!"

In a matter of seconds, funnel clouds were reported all over the

FIG. 8-1 Amateur emergency nets swung into action when this killer tornado flattened a mile-wide swath through Wichita Falls, Texas, in April 1979. (*Wichita Falls Times & Record* photo.)

Birmingham area, including twin funnels almost directly over the downtown section. The reports from the two amateurs were relayed immediately to officials. Warnings were issued at once. As more reports came in by amateur radio, the headquarters operator relayed the information to the National Weather Service on a countywide civil defènse net. Other emergency nets swung into action. The headquarters station jumped from one frequency to another, funneling traffic to Red Cross, civil defense, law enforcement, and utility officials. For more than 24 hours, amateur radio was the only means of communications out of the area hardest hit by the twisters. More than 100 amateurs were involved in just a 2-day period.

As these examples illustrate, amateur radio operators serve to expedite, clarify, and simplify communications. The reasons for them can run the gamut from mere convenience to improving the efficiency

FIG. 8-2 Volunteers who keep things running smoothly at community events such as walk-a-thons keep themselves—and their equipment—prepared for the next emergency.

of emergency communications services when time, accuracy, and maximum volume of traffic are all critical. As amateur radio has developed into many specialty areas of interest, each has produced its own customs and procedures. The newcomer should learn the tricks of the trade to participate successfully in any of the popular amateur activities. Having an experienced ham on hand to provide advice and assistance is a painless way to acquire these new skills, but just listening to the various types of activity on the amateur bands can eliminate the mysteries, too. Such helpful books as ARRL's *Operating Manual* distill the fruits of years of experience by the top experts in their fields.

This chapter outlines the basics pertaining to the most popular ham activities in sufficient detail to get a new ham started.

Contesting

The objective in just about any amateur radio contest is to make the maximum number of valid contacts in the shortest possible time. This is no time for idle conversation; it's hello and goodbye, leaving the amenities for another time. One does not use two words when one will do, or a word when an abbreviation will do. The pressure builds and the sheer volume of activity can be unbelievable. It takes great skill and quick reactions to sense when a frequency is open and jump in before someone else does. The competition is intense.

Contesting does not require any special equipment, but it does demand a station laid out for comfortable, efficient operation. There is no time to push wires out of the way or to move gear around while operating. Everything needed—microphone, key, operating controls, logbook, pencils—must be within easy reach without interference from other objects. The strong signal always has an advantage, so the bigger the antenna and the better its location, the better the contesting results. Contesting makes for good operators and efficient stations!

Rules for the different contests vary, and the rules for a given contest change in detail from year to year, so it is important to read the rules for a particular contest carefully and thoroughly, taking special care to understand the scoring system so that extra-credit contacts may be noted and sought in the contest.

To maximize contest fun and improve performance, pick your bands and times of operation carefully. Look for the band with the most activity and the loudest signals. If the howling melee in the middle of the band is disconcerting, try drifting toward the edges. Make calls and exchanges of information short—and do not repeat either unless requested to do so. The new contester finds that the time between contacts quickly dwindles, and he or she will be making more in 1 hour than in 2 hours previously.

A contest contact or "exchange" usually consists of call signs, signal reports, and serial numbers. In some contests, section or state is sent in place of a serial number.

A typical CW exchange in its entirety might be as follows. W1ABC has called CQ. WØXYZ answers with DE and his call only:

DE WØXYZ
WØXYZ 579 CT W1ABC K
W1ABC R 599 MN WØXYZ

FIG. 8-3 This station, I1HAG in Italy, could be described as a contester's dream. Note the well-organized station layout that keeps the operator within arm's reach of all his gear.

Both stations identified themselves and each other, both sent a signal report and identified their states, and the whole contact took 1 minute or less. Unlike what they would have done in other forms of amateur radio contact, these operators supplied no names, addresses, or other personal information. There is no better way to improve code speed and operating skills!

The rules for the ARRL 160-meter Contest are simpler—and shorter—than some, but they are illustrative:

1. This contest will start at 2200 UTC Friday, December 2, and end at 1600 UTC Sunday, December 4. This is a 42-hour period with no limitation on operating time. CW only.
2. The contest is open to all amateurs. A QSO with an amateur in an ARRL section (see page 8, *QST*) is worth 2 points. QSOs with amateurs not in an ARRL section are worth 5 points. DX-to-DX QSOs will not count.
3. Multipliers are the 74 ARRL sections, VE8, and each foreign country worked.
4. The exchange will be the report, plus ARRL section, for those in an ARRL section. Those participants outside an ARRL section will send a report and the name of their country.
5. Competition is within the section and non-W/VE country for certificate awards. Division high scorers will have their section award endorsed with an appropriate seal. Multi-operator work is permitted with scores to be shown after single-operator listings (no certificates).
6. Contest work may be reported either on the forms available from headquarters or on a reasonable facsimile. An entry consists of the log and summary sheet. Check sheets are not mandatory.
7. Entries become the property of ARRL; none can be returned. Awards Committee decisions are final. Send an addressed stamped No. 10 envelope for appropriate entry forms. All entries must be postmarked no later than January 9 to be eligible.
8. Standard disqualification criteria apply; see January *QST*, page 85.

DX Tips

Working stations in foreign countries happily and successfully is a matter of being in the right place at the right time, which means being on the right bands when they are "open"—when conditions and time of day are right for that band to the DX area of the operator's choice. The most important key is listening—listening to find out who is on from where and listening to learn the procedures they are using.

In the early morning—before and for a short time after sunrise—80 and 40 meters are good bands on which to start listening for signals from the west. Ten and 15 meters are likely to bring in signals from the Far East and South America in the afternoon. By late afternoon, 20 meters can produce good openings to Europe and Africa; it is usually open to *somewhere* around the clock. In the evening, 40 and 20 meters are where most of the DX action will probably be, but 10, 15, and 80 meters may provide some good opportunities, too. Experience will demonstrate how the pattern

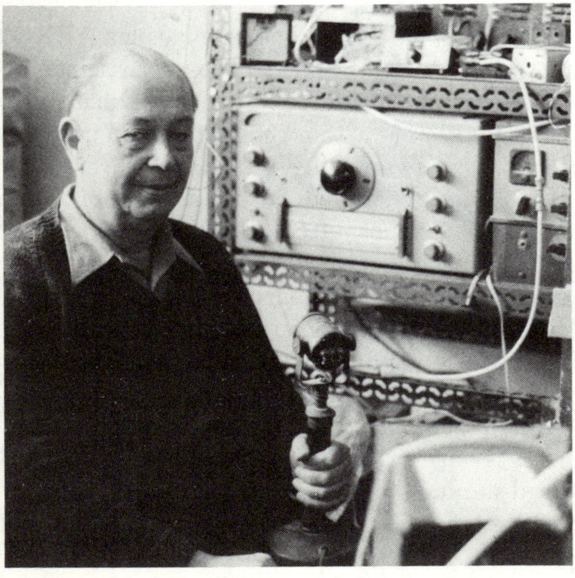

FIG. 8-4 French amateur F9OF continues one of the most exciting traditions in amateur radio—DX. Jean is an accomplished DX operator.

works for a particular location and how it varies during the sunspot cycle.

Ordinarily, a response to a CQ from an overseas station, or a call to that station when it has finished a contact, will produce results. The station is very likely to answer. But if many other U.S. stations are also calling—if the station is a DXpedition, a very rare or new DXCC country—the result can be a classic "pileup." An inexperienced operator might try all day to get through to such a station and never succeed. Most "rare" DX stations prefer to call CQ rather than answer one. They know that others want to work them, so they do not waste time answering CQs. (Unless, of course, they want to have some fun and shake up a CQer!) It is usually better for the Stateside operator to listen up and down the band for DX CQs.

The experienced DXer, if he or she is not responding to a call from a rare station, will usually make short calls, listen, make a short call again if unanswered, listen some more, and so on. Only the uninitiated waste frequency space and air time with interminable calling procedures. It is always important to listen at least briefly before making any call to avoid interfering with a QSO on a frequency already in use.

Some special techniques can help in working DX, especially if there is a lot of activity and competition on the band. Listen to a DX operator you would like to work as he or she contacts several others; you may learn how best to tailor your operating to get through. If the DX station seems to be working "tail-enders," try it; transmit your call sign quickly and briefly as another station signs off with the DX. If the DX station seems to be working only a call area or country other than your own, don't call—wait for your turn. On CW, try responding just above or below the DX station's frequency (500 Hz or so); you may be heard when others fail because of their mutual interference. If you speak the DX operator's language, try using it; you may stand out in the crowd. Vary your calling method—send just your call sign, nothing else. Try to match the DX station's code speed. Or match the code speed that successful callers have been using. None of these "trade secrets" is guaranteed to work, of course, but experts have found that any one of them can make that small but all-important difference. In any case, the best DX trick of all is: listen—then listen some more. Like contesters, the DXers with big antennas and higher power do have an advantage, but the skillful operator can often outdo the big guns without either.

U.S. amateurs sometimes do not realize—or they forget—that there are fewer hams in most other countries of the world—and hams there often lack the affluence to afford the kinds of equipment so common here. If the country is a "rare" one with particularly few amateurs, just getting on the air for a chat with a friend can be difficult. A ham in the African country of Gabon, right on the equator, writes with feeling of the problems of the radio amateur there:

> Imagine a tropical storm has just blown your two-week-old beam off the roof. Your six-year-old is down with his fourth bout of malaria. You've just killed the third poisonous snake to enter your yard in two weeks. You've worked two hours overtime, leave the office into a downpour, and your windshield wipers don't work. It's enough to make you want to escape your tropical paradise. You manage to get home, pop open a cold drink, pull up a comfortable chair and tune up the rig. You tune around the band. Ah, there's my friend, Joe. You give him a call. He comes back, "Good to hear you on, old buddy. How's it going?" Then it starts. "XYZ, XYZ, this is ABC, ABC, do you copy? Are you going to work some other stations tonight?" "A short one, a short one, please. Just a quick signal report, please." "Break, break. Info, info, break for info. Do you copy, do you copy?" Great relaxation, right?

Sought-after DX stations are well aware of their attraction for amateurs in the populous countries, and they usually try to be obliging in providing contact. But courtesy and consideration for the other operator are important on the air, too. U.S. hams should remember that amateur radio is as much a hobby and relaxation for the DX operators as it is for them. And DX operators also like to be able to pursue their own particular interests—which may not always be DX-chasing.

Operating Achievement Awards

Awards are visible evidence of an amateur's skill, proficiency, perseverance, and tenacity. Some are famous and have been around a long time. Some are much coveted. Some are difficult to qualify for;

some require less effort. One attraction of award chasing is that amateurs can work for awards at their own paces, according to their own inclinations. They compete only with themselves.

For new Novices, ARRL offers the First Contact Award; all that is necessary to qualify is a QSL card or letter certifying that a first contact took place, with the applicant's call, the call and location of the station contacted, and the frequency, date, and time of the contact.

The ARRL Rag Chewers' Club is a traditional first certificate. It requires just a solid half-hour of on-the-air conversation, and application, with details of the QSO, to ARRL.

Perhaps the most popular award is ARRL's Worked All States award. It takes some doing—Nevada, Delaware, and North Dakota can be frustratingly difficult to QSO—but it can be accomplished on the Novice bands, and often has been. A set of QSL cards confirming contacts with each of the 50 states should be submitted, with the proper application form, to ARRL Headquarters. Once the basic WAS has been achieved, a ham can go after endoresments for WAS on phone, CW, or a given band. There is even an OSCAR WAS for contacts through an orbiting satellite. Finally, there is five-band WAS, a challenging accomplishment requiring contact with a station in each of the 50 states on each of any five bands. ARRL will supply, on request, instructions and rules for WAS, a convenient application form, and a WAS U.S. wall map you can post on the shack wall as a visual record of progress.

Worked All Continents, or WAC, is sponsored by the International Amateur Radio Union (IARU) and is issued in the United States by the IARU member society, ARRL. It requires confirmation contacts with each of the six continents. All six QSOs must have been made using only one call sign from one metropolitan area or within a 25-mile radius. Additional WAC awards are available for all-SSB (phone) and all radioteletype. For those who can meet the challenges there are five- and six-band WACs. A stamped, self-addressed envelope should be enclosed for return of the QSL cards submitted. Any card with an erasure or word written over cannot be accepted.

The prestigious ARRL DX Century Club or DXCC award is available in six types—mixed, phone, CW, RTTY, 160 meters, and satellite. It is issued on confirmation of contacts with amateur sta-

FIG. 8-5 The Bicentennial Year brought excitement to hams all over the world who strived to work all 50 U.S. states. Everyone who managed to do so during that year qualified for the ARRL's special bicentennial Worked All States award.

tions in 100 or more countries as defined by the ARRL DXCC List. With more than 300 "countries" recognized, an amateur can receive additional endorsements for his or her mixed, phone, and CW DXCC certificates for submissions over 100. Detailed instructions and the DXCC Countries List are available from ARRL. The five-band DXCC award, one of DXing's greatest challenges, is awarded for confirmed contacts with 100 countries on each of five bands since January 1, 1969. The award itself is a handsome plaque with an engraved plate bearing the winner's call and the award date.

Each month, *QST*, the journal of the ARRL, lists the schedule of Morse code Qualifying Runs transmitted by its headquarters station, W1AW. Five minutes of plain text is transmitted at speeds from 10 through 40 words per minute. To qualify for a Code Proficiency Certificate, a ham must copy a solid minute correctly. Endorsement stickers are awarded for subsequent qualification at higher speeds. This certificate is issued free of charge. League membership is not a requirement for this award; in fact, an applicant need not even be a licensed amateur.

Traffic

Of all of amateur radio's specialists, none have a longer tradition or a more formalized basis of operation than the traffic handlers. Traffic traditions go back to the very beginning of organized amateur radio; the development of efficient systems and procedures for getting messages through has been the name of the traffic game. Much of the satisfaction of becoming an active traffic handler comes to a ham as he or she develops the organizational, procedural, and operating skills that mark a traffic expert—because traffic handlers are a body of amateurs who take great pride in their "professionalism."

The heart of the amateur radio traffic system is ARRL's *National Traffic System* (*NTS*) and its echelon structure of national, region, section, and local traffic nets. These meet on the air on a regular daily schedule; becoming involved can be as simple as merely "checking in." ARRL publishes a *Net Directory*, comprising a list of nets and the times and frequencies they meet. Additional information on local nets can be found in *QST* or other monthly amateur radio magazines.

Most traffic nets handle written messages in a standard form. (In fact, the FCC requires that copies of traffic be kept by the originator for at least a year.) There are four basic parts to a standard message: preamble, address, text, and signature. The preamble includes the message number, its priority or precedence, station or origin, check (number of words in the message), place of origin, time filed, and date. The address specifies the name, address, and phone number of the addressee. The text is almost always brief—20 words

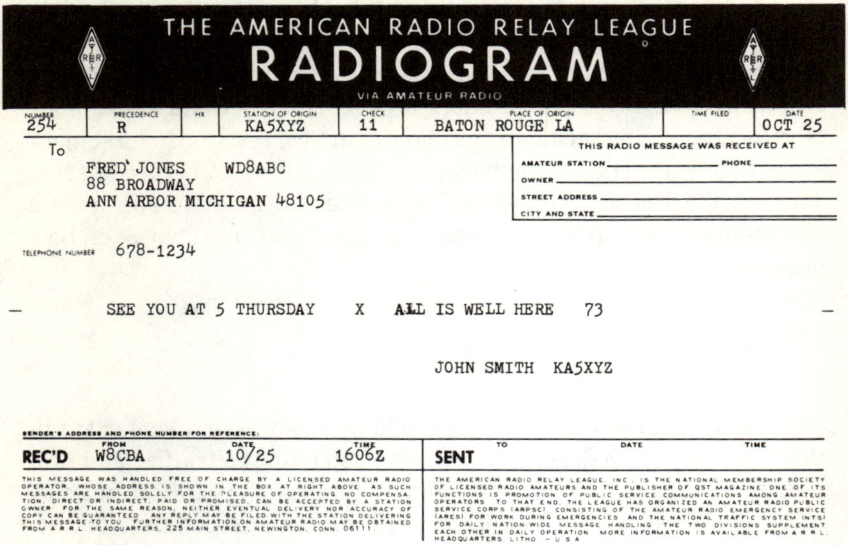

THE AMERICAN RADIO RELAY LEAGUE
RADIOGRAM
VIA AMATEUR RADIO

NUMBER	PRECEDENCE	HX	STATION OF ORIGIN	CHECK	PLACE OF ORIGIN	TIME FILED	DATE
254	R		KA5XYZ	11	BATON ROUGE LA		OCT 25

To
FRED JONES WD8ABC
88 BROADWAY
ANN ARBOR MICHIGAN 48105

THIS RADIO MESSAGE WAS RECEIVED AT
AMATEUR STATION ____ PHONE ____
OWNER ____
STREET ADDRESS ____
CITY AND STATE ____

TELEPHONE NUMBER 678-1234

SEE YOU AT 5 THURSDAY X ALL IS WELL HERE 73

JOHN SMITH KA5XYZ

SENDER'S ADDRESS AND PHONE NUMBER FOR REFERENCE:

	FROM	DATE	TIME
REC'D	W8CBA	10/25	1606Z
SENT	TO	DATE	TIME

THIS MESSAGE WAS HANDLED FREE OF CHARGE BY A LICENSED AMATEUR RADIO OPERATOR, WHOSE ADDRESS IS SHOWN IN THE BOX AT RIGHT ABOVE. AS SUCH MESSAGES ARE HANDLED SOLELY FOR THE PLEASURE OF OPERATING, NO COMPENSATION, DIRECT OR INDIRECT, PAID OR PROMISED, CAN BE ACCEPTED BY A STATION OWNER FOR THE SAME REASON. NEITHER EVENTUAL DELIVERY NOR ACCURACY OF COPY CAN BE GUARANTEED. ANY REPLY MAY BE FILED WITH THE STATION DELIVERING THIS MESSAGE TO YOU. FURTHER INFORMATION ON AMATEUR RADIO MAY BE OBTAINED FROM A.R.R.L. HEADQUARTERS, 225 MAIN STREET, NEWINGTON, CONN. 06111

THE AMERICAN RADIO RELAY LEAGUE, INC., IS THE NATIONAL MEMBERSHIP SOCIETY OF LICENSED RADIO AMATEURS AND THE PUBLISHER OF QST MAGAZINE. ONE OF ITS FUNCTIONS IS PROMOTION OF PUBLIC SERVICE COMMUNICATIONS AMONG AMATEUR OPERATORS. TO THAT END, THE LEAGUE HAS ORGANIZED AN AMATEUR RADIO PUBLIC SERVICE CORPS (ARPSC), CONSISTING OF THE AMATEUR RADIO EMERGENCY SERVICE (ARES) FOR WORK DURING EMERGENCIES, AND THE NATIONAL TRAFFIC SYSTEM (NTS) FOR DAILY NATION-WIDE MESSAGE HANDLING. THE TWO DIVISIONS SUPPLEMENT EACH OTHER IN DAILY OPERATION. MORE INFORMATION IS AVAILABLE FROM A.R.R.L. HEADQUARTERS. LITHO - U.S.A.

FIG. 8-6 Hundreds of messages like this one make their way through efficient National Traffic System nets that are run entirely by volunteer hams.

or less—with no punctuation. Sentences are separated by the letter X.

To check into a CW net, send a letter, perhaps M, to get the Net Control's attention. He or she will acknowledge by repeating the letter, and that acknowledgment is your invitation to check in. The procedure might go something like this:

> The NCS (Net Control Station) calls the net together:
> CN DE K1EIR QNI K (Connecticut Net from K1EIR. Stations wishing to check in, go ahead.)
> M (A station wishes to catch NCS attention.)
> M (NCS K1EIR acknowledges.)
> K1EIR DE WA1VMC GE QTC NEWINGTON 1 $\overline{\text{AR}}$
> (Good evening. I have one message for Newington.)

As soon as possible, the NCS will move WA1VMC and a receiving station off the net frequency to pass the traffic. When WA1VMC returns, he or she will let the NCS know by sending his or her call sign suffix, VMC. If the message is destined for a state in another region, the operator originating it would specify QTC 1 THRU.

At the conclusion of its NTS journey, a message will usually be received by an amateur within toll-free telephone calling distance of the addressee. Several points are important when a message is delivered by telephone. The message is introduced as a greeting via amateur radio—a sudden telegram-like message from a distant friend or relative can cause alarm! The receiving operator explains a bit about amateur radio and how the message was handled through the hams' own National Traffic System. Finally, the receiving operator offers to originate a reply message if the addressee so wishes. The message is thus efficiently delivered, and amateur radio gains a friend through another NTS job well done.

Messages are processed by a net according to their priority, specified by the originator of the message. "Emergency" is a message classified as having life-or-death urgency. "Priority," the next-lower classification, is assigned to official or important messages with a specific delivery time limit. "Inquiry" is used for a message inquiring about the health or welfare of someone in a disaster area. "Routine" is applied to all other messages; it is used for most traffic in ordinary circumstances.

As a readiness exercise and check of coordination between NTS and the Amateur Radio Emergency Corps, the several thousand NTS members participate every October in a Simulated Emergency Test, or SET, sponsored and coodinated by ARRL. In the SET, participating amateurs explore their strong points and limitations in providing emergency communications; they upgrade their expertise through the use of standard operating procedures under the conditions of simulated emergency; and they demonstrate to the mass media and agencies served (such as the Red Cross) the usefulness of amateur radio as an emergency communications service.

VHF/UHF—The World Above 50 MHz

"Homebrew" (homemade) gear is still the order of the day for the serious upper-frequency user in amateur radio. This is the world of the experimenter and today's amateur radio pioneers. Most neophytes will buy or build equipment for one VHF/UHF band at a time because good multiband gear is rare and relatively expensive. Antennas for the higher bands are smaller in general than those used on the traditional high-frequency amateur bands, but more complex than the familiar dipole or vertical. Successful operation on the upper bands also requires a good understanding of their propagation characteristics. Contesting above 50 MHz is often referred to as "hilltopping," because getting up high is a real advantage on these line-of-sight bands. Equipment used ranges all the way from miniature hand-held units to very sophisticated rigs and elaborate antenna systems.

Two-meter FM, extending from 144.1 to 148 MHz, is by all odds the most popular amateur radio activity today, making 2 meters the most popular ham band. At 144 MHz, it is well above the 50-MHz VHF/UHF dividing line, but it really is a special band. Its users almost invariably turn to the small commercial transceiver units, either hand-held or car-mounted, and they use them for practical, local communications. At frequencies other than those in the re-

FIG. 8-7 After just a couple of weekends of building time, these components will fit into an amateur's hand as a 2-meter transceiver.

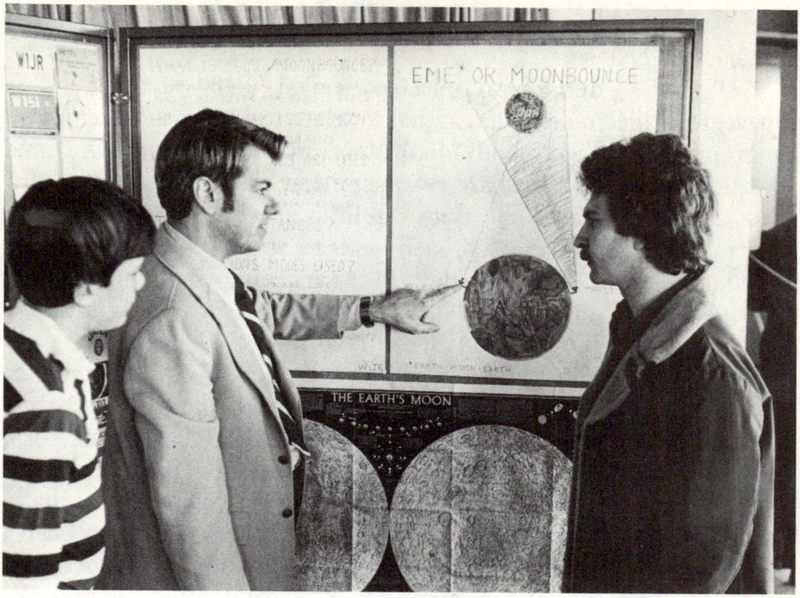

FIG. 8-8 Well-known VHF/UHF experimenter Joe Reisert, W1JR, explains moonbounce techniques to a newcomer. Sending amateur radio signals to the moon and back is just one of the new frontiers that hams are involved with. (Photo by Joel Kleinman, WA1ZUY.)

peater subbands, however, 2 meters does come in for exotic uses on SSB and CW—moonbounce, meteor scatter, and OSCAR.

The other popular VHF/UHF bands are 6 meters (50 MHz), 1¼ meters (220 MHz), 70 centimeters (420 to 450 MHz), and 23 and 13 centimeters.

Space Communications through OSCAR

The percentage of OSCAR users among the total amateur population is still relatively small, and it is probably safe to say that the reason for this lies in psychological barriers rather than physical ones. Most hams have (or have ready access to) equipment to work through OSCAR satellites, and the special techniques required can be mastered rather easily by anyone holding a Technician or higher-class license.

OSCARs 7 and 8 both contain so-called "2-10 transponders" or repeaters, which means that signals *to* the satellites are transmitted by ground stations around 145.9 MHz, in the 2-meter band, and signals *from* the satellites are retransmitted by OSCAR around 29.5 MHz, in the 10-meter band. Both satellites are in a sun-synchronous, polar orbit, which means that each tends to be overhead and in range at about the same time each day (in the morning and evening, local time). At any given time, each satellite's coverage area extends up to 2500 miles from the point on the ground directly beneath it.

An easy way to build confidence in using OSCAR is simply to listen. All you need is a receiver for the 10-meter band and information on when to listen, which is available from ARRL headquarters. If you tune to the downlink frequency on 10 meters 15 or 20 minutes before a pass scheduled over a nearby city, signals will gradually become audible as the satellite approaches, reach a peak of intensity when it is nearly overhead, and then diminish as it recedes over the horizon. A little practice at tuning back and forth around the frequency during the period of the pass will have OSCAR QSOs coming in loud and clear. It is quite an experience the first few times. First there is an empty, silent band; then suddenly activity begins, signals explode, and there is a barrage of conversation. Then, as suddenly, the band is quiet again. Regular OSCAR users frequently add a 10-meter preamplifier to their receiver systems to increase sensitivity on the OSCAR frequency and reduce noise and interference levels. Soon the Doppler effect and fading caused by Faraday rotation, phenomena characteristic of space signals, will become familiar to the beginning OSCAR listener.

Experienced OSCAR-types say that almost any rig capable of generating a few watts of radio frequency energy at the appropriate frequencies on 2 meters is suitable for an OSCAR uplink transmitter. Old 2-meter AM and CW transmitters, converted commercial FM gear, and commercially available transceivers and transverters will work fine. Ten watts or so is plenty; high power is not needed for OSCAR work. In fact, excessive power can even damage the satellite. Only enough power and antenna gain to get through should be used. For reasons of efficiency and availability of equipment, CW

FIG. 8-9 Launched into orbit in 1974, AMSAT-OSCAR 7 provided amateurs with a reliable—and fascinating—means of communicating with others up to 5000 miles away.

seems to be the most popular mode on OSCAR. Even a 2-meter FM unit with the proper crystals, keyed for CW, can be an acceptable OSCAR transmitter.

The experienced OSCAR user has a commercial 11-element 2-meter beam antenna, mounted with vertical polarization on a horizontal mast, and equipped with two rotors and controls for elevation and azimuth. A directional antenna must, of course, be pointed at the sky to pick up OSCAR, and a tracking mechanism makes for maximum signal strength and maximum access during a pass. The neophyte, however, will find that he can get very satisfactory initial results with an omnidirectional fixed dipole.

OSCARs 7 and 8 have beacons that can be heard at 29.502 and 29.402 MHz, respectively, in the 10-meter band. The beacons transmit CW signals reporting the condition of various internal systems. Copying these telemetry signals is not only interesting; it is good practice.

In addition to the 2-10 transponders (satellite operating Mode A), both satellites also carry secondary transponders. OSCAR 7's Mode B has its uplink on 432.125 to 432.175 MHz and its downlink on

145.975 to 145.925 MHz. OSCAR 8's Mode J has its uplink on 145.900 to 146.000 MHz and its downlink on 435.100 to 435.200 MHz. In other words, the uplink and downlink bands are reversed for the two satellites.

Satellite operation is full duplex, which means that the satellite transmits and receives simultaneously without need for switchover. Thus, operators can monitor their own signals as retransmitted by OSCAR; they can know exactly how their signals sound and what the receiving frequencies are. They can call CQ on those frequencies and expect a quick response. Contacts on OSCARs 7 and 8 tend to be very brief because of the short duration of any one pass.

One advantage of using the alternate transponders—Mode B on OSCAR 7 and Mode J on OSCAR 8—is that contacts can be made with very low power. One ham reported: "The results really amazed me. Three watts output on phone brought 29 contacts in 14 states and six countries in just three days!" Other comments: "We were able to contact each other with as little as 2 watts." "I got an acknowledgment to my call while I was unintentionally running only 50 milliwatts."

The organization responsible for the design and launch of all recent OSCARs is AMSAT, the Radio Amateur Satellite Corporation. AMSAT members worldwide receive a magazine that reports on all operating aspects of the satellites as well as technical developments. Orbital data and satellite operating schedules are published regularly. Amateurs—members or nonmembers—who send OSCAR 7 telemetry reports to AMSAT (P.O. Box 27, Washington, DC 20044) receive special AMSAT QSL acknowledgment cards. AMSAT also offers two awards: the Satellite Communicators Club for just one confirmed two-way contact through a satellite, and an OSCAR Achievement Award for performance at more advanced levels. ARRL has its own Satellite DXCC, OSCAR Worked All States and Satellite DX Achievement awards. Send OSCAR 8 telemetry reports to ARRL Headquarters.

QRP Operating

Working with low power, the lower the better, has been attracting more and more amateurs. It has something of the challenge of playing golf with just one club, or sports car rallying, where skill and timing are far more important to success than speed. The emphasis in QRP is certainly on operating skill and experience.

Basically, QRP operators cannot just jump into a band. They have to do a lot of listening and a good deal of creeping around the edges of active frequencies. They soon learn just the right timing for hitting the key with a CQ call—extremely important in working with low power. The QRP signal must be clean and stable. VFO frequency control, rather than crystal control, is preferred because it allows much greater flexibility in operating. A good trick is to look for calling stations with particularly strong signals, on the assumption that their receiver sensitivity is also likely to be good for picking up low-power signals. A good antenna is highly desirable, and low-loss transmission lines improve performance. QRP gear is attractive to the ham with a small budget because it is relatively inexpensive.

As QRP has grown in popularity, specific HF-band frequencies have been set aside by informal agreement for low-power contacts

FIG. 8-10 Recognizing the role amateurs play in emergency situations, states and provinces issue distinctive call sign license plates to hams. When a car with such plates arrives at the scene of an emergency, police recognize the driver as a source of assistance.

and net operations. On 20 meters, the frequency is 14,060 kHz; on 40, 7030 kHz; and on 80, 3506 kHz. Checking into one of the QRP nets is an excellent way to get into low-power operating. There is also a regular schedule for QRP on OSCAR.

Identifying Hams

Identifying and locating Al, W1ABC, with whom one has just finished a QSO, is a simple matter. Amateur radio has its own directory. In fact, it has two directories, one for the United States and one for the rest of the world. The two volumes list nearly every licensed radio amateur alphabetically by call sign for each U.S. call area and each foreign country. Each listing gives the amateur's full name, address and call sign. The directory is entitled the *Callbook* and is published annually, with three quarterly supplements, by Radio Amateur Callbook, Inc., 925 Sherwood Drive, Lake Bluff, Illinois 60044. The *Callbook* also contains other useful information, such as worldwide QSL Bureaus, world amateur prefixes and map, standard time charts, and the ARRL DXCC list.

9

Hamming Around the World

It was January when Dave Bell, W6AQ, first started thinking about a DXpedition to the Orient. By March he had decided that the end of October would be the ideal time to go. Here's his diary account of subsequent events:

March 27 . . . I searched through my QSL cards for a DX station to "borrow" so I would not have to take a lot of equipment. Decided on CR9AJ (Macao) and VS6DR (Hong Kong) as my prospective innocents aboard.

March 30 . . . Called ARRL for advice. Their international expert told me Macao might be difficult but that we had a reciprocal licensing agreement with Hong Kong, and if all else failed, there were lots of islands in the Pacific.

April 19 . . . After two weeks of head scratching, I wrote to

Torres, CR9AJ, and as subtly as possible attempted to thrust myself upon him for the last weekend in October.

May 6 . . . Having heard nothing from CR9AJ, I wrote another letter with a copy of the first. Some knowledgeable DXer friends had expressed skepticism about mail delivery to Macao.

July 7 . . . At a meeting of the Southern California DX Club, I was urged to forget about Macao, since I hadn't heard from Torres. Try Hong Kong, I was told.

July 8 . . . Wrote Phil, VS6DR, with the same sort of "borrow your shack" letter I'd written to Torres.

July 12 . . . At ARRL's suggestion, I wrote David, 9V1RH (Singapore), for advice and assistance since he knows everything about everything in his part of the world. Specifically, I wanted help with Macao since I had zeroed in on that as my prime prospect.

July 19 . . . Heard from 9V1RH, proving that it is possible to send and receive transpacific mail in less than an eternity. David's first paragraph said, "Concerning a license in Macao, frankly, I don't like your chances." If that weren't enough, he let me know that in Singapore one must have a permanent address to obtain a 9V1 call, and he suspected that Hong Kong, even with a reciprocal agreement, might be the same.

August 2 . . . A letter from Phil, VS6DR, who had been out of town, reported that his antennas were down and he was temporarily off the air.

August 3 . . . Sent letter to Hong Kong version of the FCC, enclosing all the required information to obtain a reciprocal license. Ham licenses there are handled by the post office.

September 6 . . . This day I take back almost everything bad I ever said about post offices. I heard from CR9AJ. It had taken Torres just slightly over four months to find out that I couldn't get my own CR9 call, but it would be okay for me to use his call and his station. His letter was warm and welcoming like a sea breeze on a smoggy day!

October 13 . . . Letter from the Hong Kong FCC stated that I

FIG. 9-1 Torres, an amateur on the island of Macao, shows off the local sights to Dave Bell, a California ham who took a "radio vacation" to the island.

can get a license if I stay for more than three months. However (like Macao), they have no objection to my operating an existing station if it's okay with the licensee. Maybe next time.

October 26 . . . My wife and I leave LA International. It's the first leg on the way to Macao.

October 29 . . . Standing on Guia Hill at 5 A.M., watching the ancient junks heading out to sea was an incredible experience. I'll not forget it soon. But first things first. It's time for Torres to introduce me to his station. 10-15-20, here I come!

Now, admittedly, this amateur's experience was not typical of the situation facing most U.S. hams wishing to operate in a foreign country. But it does illustrate the fact that it takes some planning and arranging. Amateur radio is a regulated activity—more so in some countries than others. And each has its own requirements for its own citizens and for visitors.

The United States has executed Reciprocal Licensing Agreements for amateur radio with a number of countries. These agreements, in effect, represent an exchange of courtesies to make it possible for amateurs from one country to operate while visiting others. These agreements are by no means universal, and where they do exist they must be consistent with each country's own rules. Thus, the requirements vary widely. The ARRL has separate instruction sheets as a guide to the requirements in almost 100 countries; specific details on each country of interest are available on request to any amateur planning to travel abroad.

The U.S. government has traditionally been a strong supporter of amateur radio. It has always believed that radio amateurs represent a valuable national resource for emergency communications, for technical and scientific training, and for building international goodwill. U.S. delegations to international frequency-allocation conferences have been in the forefront of proposals on behalf of the world amateur radio community. As a consequence, hams in this country have a relatively favored status and liberal operating privileges. Over the years, the American Radio Relay League, the hams' own organization, has been an effective spokesman for amateur interests and has worked closely with our regulatory agencies.

Amateur Radio in the Developing Nations

In many other countries of the world, telephone and radio communications are government operations; some governments view radio amateurs as potential competitors and restrict their activities accordingly. Other controlled societies are concerned about communications in general between their citizens and the outside world. To them, amateur radio solely for the pleasure and enjoyment of its practitioners would be frivolous; in any such activity, the national interest must in some way be served directly. As a result, getting an amateur license in some countries can be a lengthy and difficult process to start with, and operations that are permitted are restricted.

Economics is also an important factor. Few countries have populations as generally affluent as that of the United States. The

FIG. 9-2 This modern building was the site of a conference in 1979 that reviewed and, in some cases, reallocated, portions of the international frequency allocations. The Amateur Radio Service gained several new bands at the conference, which was held at Geneva, Switzerland.

kinds of amateur station equipment U.S. hams take for granted are simply unavailable in much of the rest of the world.

U.S. citizens and other hams in Western countries often take amateur radio for granted. When they describe their hobby to friends and acquaintances, they assume that its sanctity as a privilege is understood. But that is by no means the case in many countries. Indeed, there are languages in which there is no word or phrase for the concept of "amateur radio." Some national amateur societies have only a handful of members—the entire ham population of the country! The Mongolian People's Republic, for example, is home to a total of five licensed amateurs. That certainly makes JT1 a rare DX contact!

As a government official described the situation in his emerging African nation:

> You must remember that the African does not view his life in the same way as a Westerner. He does not see Africa as one unified continent, but rather as a great, large land of many differences:

FIG. 9-3 In Africa, amateur radio is a means of gaining technical skill, rather than a hobby. These newly licensed amateurs will use the hands-on experience that amateur radio provides to help their nation's progress. (Photo courtesy Botswana Amateur Radio Society.)

climate, geography, language and, most of all, problems. For so many Africans on this grand continent, life means barest subsistence, and a meaningful life is one in which one can meet the needs of one's family—the vital core of African life. The very concept of "hobby" is foreign to us. What does this mean, "leisure time"? Amateur radio here is not viewed as a hobby. To enlightened governments it is a means of creating a corps of self-trained technical experts so vital to a developing nation's telecommunications and technical growth. Too, it is a perfect source for an emergency communications network. And, of course, it is a system by which a country's youth can be occupied and directed. That is why African schools are interested in promoting amateur radio.

The tradition of active communication—the spoken word, drums, and music—runs deep in African history and culture, so direct communication is of special importance to the African people. A Kikuyu tribesman, shown a book about amateur radio, said: "Yes, I know what it says there. But now tell me with your own voice and I will believe you." That tradition makes the person-to-person nature of amateur radio notably appealing to Africans. To give encouragement to the development of amateur radio in Africa—and elsewhere—the ARRL sponsors a program for distribution of a basic transmitter and receiver kit to clubs and individuals in these countries. Clubs and individual hams in the United States have underwritten many of these donations.

Operating While Abroad

For U.S. amateurs traveling abroad, the fact that amateur radio is indeed an international fraternity will soon assume very real proportions. Their ham call signs will open many doors, establish them immediately as warmly welcomed friends, and provide them with all kinds of assistance. The hospitality they will receive will be a far cry from that accorded the ordinary visitor. If they have had previous contact with the host hams on the air, the welcome will be especially warm, but previous contact is hardly necessary to guarantee a hospitable reception.

New hams with no personal contacts in countries they plan to visit have several sources for such contacts. They can get names and addresses from other hams or fellow radio club members in their communities who do have overseas contacts. They can write the national societies or radio clubs in the countries they plan to visit, and they can cull names and addresses in particular cities from the DX *Callbook*. Or they can call or write the Membership Services Department of the ARRL. ARRL's MSD personnel can provide the addresses of national societies and leading radio clubs, with details on operating requirements in their respective countries, and perhaps even an individual name or two. ARRL can also supply general advice based on extensive experience and work with amateurs in countries around the world.

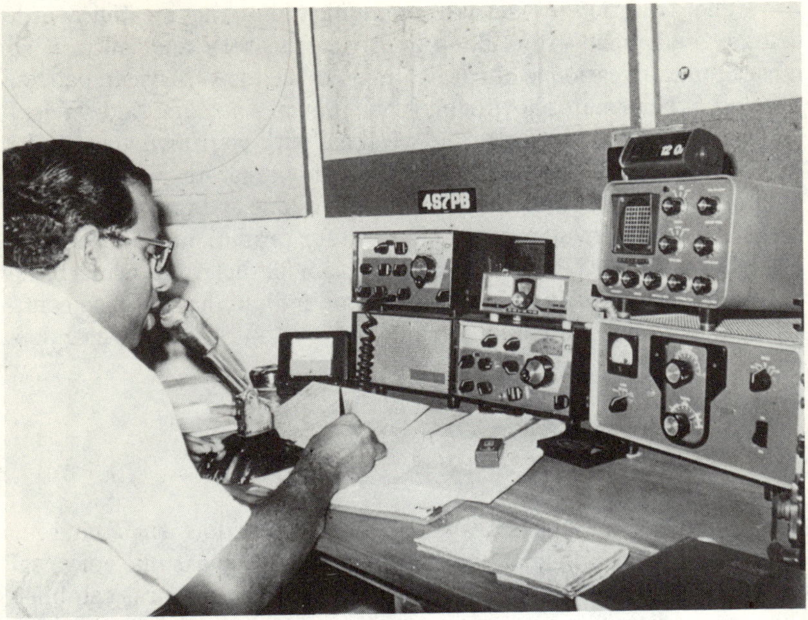

FIG. 9-4 One of the few amateurs on the island of Sri Lanka (formerly Ceylon), 4S7PB is a sought-after contact from amateurs all over the world. Although this ham is fortunate enough to have modern equipment, amateurs in much of the world must rely on older, less-reliable gear.

U.S. radio amateurs are free to exchange messages in this country on behalf of others—messages known as "third-party traffic." In many other countries, however, third-party traffic is not allowed. As mentioned earlier, some governments view such amateur service as competitive with their national telephone and other communication services; they permit only routine contacts between their amateurs and those overseas.

The countries that permit third-party traffic with the United States are: Argentina, Bolivia, Brazil, Canada, Chile, Colombia, Costa Rica, Cuba, Dominican Republic, Ecuador, El Salvador, Ghana, Guatemala, Guyana, Haiti, Honduras, Israel, Jamaica, Jordan, Liberia, Mexico, Nicaragua, Panama, Paraguay, Peru, Trinidad and Tobago, Uruguay, and Venezuela. Amateurs should check with ARRL for late-breaking changes in this list.

10

Over the Horizon

Where are the new frontiers of amateur radio? Some hams are already exploring them. Computers? More than 3000 enthusiasts attended a recent amateur radio computer hobbyist convention. The VHF and UHF frequencies? Many amateur publications now feature regular departments on what a lot of hams still consider to be an exotic subject. In the ARRL's *QST*, the monthly column "The World Above 50 MHz" has featured the science of making contacts via meteor scatter—amateur radio signals bouncing off meteors and other space debris! Satellite communication? Hams in more than 100 countries have used the OSCAR amateur satellites.

Ham Computers Now

A microprocessor is basically a data-processing unit on a single integrated-circuit chip without the various devices that would make it a complete microcomputer. Microprocessors have already found application in a wide variety of products—calculators, typewriters, cash registers, electric stoves, microwave ovens, navigation systems, industrial controls, and even automobiles. Their influence on our lives in just beginning to be felt; what the future will bring in the coming computer age is difficult to predict, except that everyone will enjoy the benefits of computers in many ways.

Because the microprocessor lends itself so ideally to small-scale control systems, it has some intriguing possibilities for applications in amateur radio. Microprocessors may be used for repeater control operations, satellite antenna tracking, contest logging, and Teletype demodulation or word storage. Future uses of microprocessors include storing and routing messages in traffic networks, and enhancement of slow-scan television images.

In essence, a microprocessor has the ability to recognize a sequence of simple control statements or instructions that will cause the electronic circuitry to perform a task. That task can be completely fixed or it can adapt itself to an anticipated set of possible conditions. Most significantly, the ultimate capabilities of a microprocessor-based system are determined by the thought and ingenuity put into the control sequence or program (known as "software," as opposed to the computer electronics and interface devices or "hardware").

The most basic piece of information a computer works with is called a "bit." A bit may have only one of two possible values. These values may be represented as "on" or "off," as "true" or "false," as "5 volts" or "0 volts," or even as "key down" or "key up." All the information processed by a computer is in the form of combinations of bits, or "words." Each word is normally at least 4 bits long. Since a microprocessor can work only with numbers, it uses numbers to represent characters; the computer display converts numbers to characters according to its program.

A microcomputer has four basic elements: inputs, a microprocessor—known as the central processing unit or CPU—memory, and outputs. The inputs and outputs interface the microcomputer with the outside world. An input, for example, may be a telegraph key, a typewriter keyboard, or a voltage. The input signal, whatever its form, must be converted to numbers composed in turn of bits. The output performs the reverse function; it takes numbers produced by

FIG. 10-1 This type of silent, easy-to-use video terminal demonstrates how well microprocessors and amateur radio go together. This microprocessor-controlled machine will send and receive three types of codes: Morse, Baudot, and ASCII.

the computer and converts them into teleprinter copy or the precise rotation of an antenna. The input and output process is commonly referred to as I/O.

The memory component of a microcomputer normally performs two distinct tasks: it holds or "remembers" the series of instructions comprising the program, and it remembers numbers produced by the CPU that will be needed for subsequent calculations. The first task is referred to as program memory; the second, as data memory. A typical memory bank can store several thousand bits of information.

The CPU is the component of the microcomputer that makes decisions and performs arithmetic computations. It can "read" a number from memory by specifying the location or "address" at which that number is stored. In most cases it can also "write" numbers into memory. The CPU is, of course, the heart of the computer. Most current microprocessors can recognize from 30 to 100 different instructions—including input–output instructions, add instructions, and conditional jump instructions. Conditional jump instructions are pivotal; they make the microcomputer a powerful tool because they permit the CPU to make a decision. For example, a conditional jump instruction may tell the CPU to take the content of the next memory location and incorporate it in the calculation if a number in a particular location is zero; if the number is not zero, the conditional jump instruction orders the CPU to add 2 to the calculation. In other words, the instruction can cause the CPU to jump to a different part of the program if the stored number is zero.

A microprocessor is designed to follow a limited set of instructions, each directing it to perform a specific task. A list of instructions to perform a particular job or combination of tasks, such as controlling an amateur repeater, is called a program. A program must be "written" in language a computer can understand. This "machine" language used differs among equipment of different computer manufacturers. What makes computer programming such a specialized skill is that a lot of machine language is necessary for the compilation of a complete program. A simple program may require 400 separate instructions, and a large program may incorporate 1000 to 2000 instructions. Not only does it take a lot of time to write such a program, it also takes a lot of time to "debug" it—to correct programming mistakes and make the program work properly.

Microprocessors are available in 4-, 8-, 12- and 16-bit "word" capabilities. Most control applications can be handled easily with words of 4 bits; other data and communication applications are easier with CPUs capable of processing words 8 or more bits in length. Most amateur computer users employ 8-bit units.

A microcomputer system for amateur radio use can be as simple as a microprocessor, a keyboard, and an output display device, but six components are usually considered minimal. In addition to the three basics, these include additional memory, a permanent program-storage unit, and special devices for interfacing the computer with station equipment. Keyboards and displays are often available as integral units. For program storage, the trend is toward floppy-disk storage systems. Like radio equipment itself, there is almost no limit to what can be spent on computer equipment—but an 8-bit microcomputer, display, and keyboard in kit form can cost as little as $150.

An Indiana amateur radio club converted its local repeater to autopatch (telephone system linkage) with microprocessor control. In addition to controlling the repeater identification and time-out functions, the microprocessor also handles the autopatch-access code, audio switching, and Touch-Tone to dial-pulse conversion. When an amateur's Touch-Tone signal arrives over 2-meter FM, the microprocessor recognizes it, responds with an R in Morse code (for Roger), and initiates the telephone call. The microprocessor also recognizes when the universal emergency number, 911, is entered and automatically dials the local Crime Alert telephone number. The microprocessor program was written by a club member, using approximately 800 eight-bit words of program storage and 100 words of data storage.

The Wide Open Spaces of VHF/UHF

With the crowding occasionally found on the most popular amateur HF bands and on 2 meters, many hams would probably cast their votes for the VHF and UHF frequencies—and above—as the next big frontier for amateur radio. And indeed there is much to support that feeling.

Perhaps the most attractive feature of the upper frequencies is one many hams do not fully appreciate—there is a lot of spectrum space there, much of it unused. The amateur HF bands, from 80 through 10 meters, offer a combined total of 3.3 MHz. In the VHF bands, from 50 through 225 MHz, amateurs have access to a total of 13 MHz worth of frequencies. The UHF 420-MHz band offers 30 MHz, and above 1000 MHz there is 2220 MHz of spectrum available to amateurs—there is a tremendous natural resource up there!

One reason there has not been a great rush to these frequencies is the general belief among amateurs that they are "line of sight," limited to local, short-range use. That is not entirely true. All radio waves, in fact, travel in straight lines unless forced to do otherwise. VHF waves are in some respects less easily reflected by atmospheric layers than waves of lower frequency, but there are many ways through which VHF/UHF wave energy can be reflected, refracted, or scattered to extend its range. Moonbounce, meteor scatter, and satellite communication are just three examples. VHFers consider

FIG. 10-2 It's not everyone's idea of a desirable addition to one's property, but this moonbounce array in Portland, Oregon, is capable of transmitting a signal to the moon and back. Amateur contacts of more than 11,000 miles have been made using antenna arrays similar to this one.

exotic applications of propagation theory the big challenge of their specialty.

Reflection of radio waves in the tropospheric layer of the atmosphere is the most common form of VHF DX propagation. Weather maps are excellent indicators of such DX possibilities because tropospheric propagation is the result of changes in the refractive index of the atmosphere at the boundary between air masses of differing temperature and humidity. The effect is most prevalent at distances under 150 miles, but it may extend much farther. Tropospheric bending is most common in the fair, calm weather of the warmer months, although it can occur at any season. A phenomenon known as "ducting" may produce such extreme DX as contacts on 144, 220, or 420 MHz between California and Hawaii or between New England and Florida.

Exploration of the world above 50 MHz actually began in the early days of radio, but the big impetus came following the accelerated technical developments of World War II. The true frontier for hams is now at the frequencies above 1000 MHz. DX contacts already achieved are indicative of the possibilites that exist. Reflection from the moon—moonbounce—shows great promise. Any two points on the earth's surface where the moon can be seen simultaneously are within moonbounce range of each other.

One of the earliest DX propagation modes discovered by upper-frequency ham enthusiasts was sporadic-E skip. It results from the reflection of VHF waves by dense patches of ionization in the E region of the ionosphere, roughly 50 miles above the earth. The causes of sporadic-E are still not completely understood and its occurrence is predictable only in a general way, but its effect are well known to generations of VHFers. Contact distance can extend to 1200 miles or more. E-layer propagation is most common in the months of May, June, and July. E openings during the peak month of June may last for several hours.

Auroral propagation involves the reflection of VHF waves from the auroral curtain in the northern skies, usually at acute angles. Although it is most common at 50 and 144 MHz, some auroral work has been done on 220 and 420 MHz, and it may eventually become feasible above 1000 MHz if very large antenna arrays are used. Scientific investigations with very high power and large antennas have produced auroral returns at frequencies of several thousand MHz. The reflecting properties of aurora vary so rapidly that voice communication via aurora is often unintelligible; CW is the most effective mode of operation for auroral work. Auroral DX has been worked as far south as 30 degrees in the southeastern United States, but the more northerly latitudes, especially in the Northeast, are much more productive. Auroras follow seasonal patterns, being most common around the equinoxes in March and September. Much is yet to be learned about auroral propagation.

Meteor scatter is one of the more esoteric forms of upper-frequency DX currently pursued by amateurs. Meteors constantly enter the earth's atmosphere from outer space and burn up rapidly in the E region of the ionosphere as a result of friction. In the process, a cylinder of dense ionization is formed as a trail behind the meteor. This can produce a burst of reflected radio signal—perhaps only a few words of understandable voice from a station up to 1200 miles away.

FIG. 10-3 The frontiers of VHF and UHF are constantly being expanded as amateurs experiment with new equipment. This "antenna farm" features an array for 432 MHz, a 220-MHz system, and a surplus dish for 1296 MHz. Note the truck, used for operating the sophisticated station.

Bursts of up to 1 minute or even more of continuous signal may occasionally be heard during major meteor showers. Meteor showers throughout the year are fairly predictable, so that prior arrangements can usually be made for contact attempts.

Lunar communication has been a longtime dream of experimentally inclined amateurs, and successful use of the earth–moon–earth

(EME) path is becoming increasingly commonplace. Some of the first results were achieved as recently as the early 1950s on 2 meters. Some communication via the moon has been carried out over long distances on each of the VHF/UHF bands, but only with very sophisticated equipment. Moonbounce remains a challenge to the more advanced amateur with access to the necessary technical resources. The maximum legal power must be used in addition to sensitive receivers and elaborate antenna systems capable of being aimed and controlled accurately. An intriguing prospect for practical lunar communication may lie in the possibility of an amateur repeater on the moon at some time in the future!

A New Era for OSCAR

With the advent of the first Phase III amateur satellite, amateur communications via orbiting spacecraft will move into a new era. The principal drawback to all previous, low-altitude, OSCAR satellites is their limited availability, less than 2 hours a day for most hams.

To achieve maximum access to the satellite by the 90 percent of the amateur population residing in the northern hemiphere (without shutting out those in the southern hemisphere), an elliptical orbit with the apogee (highest point) about over the North Pole is nearly ideal. It will provide access for 10 hours or more at a time. Phase III is almost like a whole new band. With the satellite at apogee, every station in the northern hemisphere theoretically will be able to communicate with every other—a circumstance never before possible in amateur radio. Ionospheric propagation difficulties are, in effect, nonexistent. The higher altitude of the Phase III satellite extends the communications range to 11,000 miles. This is indeed the beginning of a new generation of OSCARs.

Because of the much greater distances involved, the demands on power for the Phase III satellite far exceed those for earlier satellites. The level of the downlink signal received by users depends substantially on the power produced by the satellite system itself. The basic source of energy is an array of solar panels. Physical size restrictions on a satellite that can be accommodated as a secondary payload on the launch vehicle led to some ingenious engineering design to maximize the surface area available for mounting the solar panels. The result is approximately 1 square meter of solar cells distributed about the satellite body. These will produce about 45 watts of power at the beginning of the satellite's lifetime, a level expected to degrade to about 30 watts output in 5 years, but entirely sufficient to operate all the onboard systems. During periods when the solar cells are not exposed to the sun—periods that can be lengthy in elliptical orbits—transponder operations are assured by a nickel–cadmium battery which recharges completely during the long periods of sunlight near orbit apogee. Another technological innovation in the Phase III satellite is an onboard microprocessor which controls just about every function on the spacecraft. It executes telemetry and command requirements, monitors the condition of the power and communications systems, controls timing devices, and takes corrective action as necessary. Finally, it interacts with the attitude sensors and torquing magnet to adjust orientation of the spinning satellite in space, its most demanding task. Unlike previous OSCARs, which were injected directly into orbit by their launch vehicles, this spacecraft has to be injected into final orbit by a small solid-propellant motor.

The satellite's antenna is located on the spinning axis of the spacecraft and points directly toward the earth at apogee. The attitude of the satellite remains fixed throughout the orbit. The main antenna functions effectively at apogee over the northern hemisphere, but its efficiency falls off as the satellite descends toward perigee (the orbital low point) over the southern hemisphere. To sustain high-quality signals there, the transmitter is switched automatically to an omnidirectional antenna as the satellite crosses the equator.

Narrow-Band Voice Modulation

One of the great technological breakthroughs in amateur radio was the advent of single sideband, a mode of voice transmission far more efficient than traditional AM. Yet SSB still requires considerably more frequency space than does a CW signal. Now, a new development promises yet another big advance in amateur operating technology. Called narrow-band voice modulation (NBVM), it was developed by three scientists—two of them radio amateurs—at the University of the Pacific in California. NBVM is a technique for compressing speech frequencies so that only about half the normal bandwidth is used. The benefits on the sometimes crowded amateur bands are obvious: more QSOs per megahertz. Little modification of existing equipment is required; the frequency compandor units are installed in the microphone and speaker leads in the station. The compressed voice signals can even be received with somewhat-reduced intelligibility on a conventional SSB receiver.

The frequency compandor filters the essential parts of speech and down-converts these electronically on transmission. An additional advantage of the narrower bandwidth is that less noise power competes with the signal on reception. The system is adaptable to

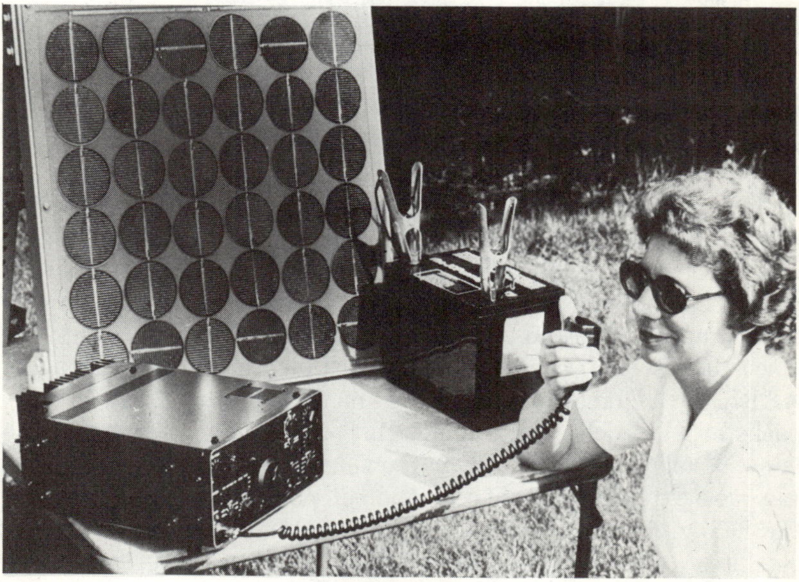

FIG. 10-4 Amateurs have traditionally been in the forefront of technological innovation. Solar cells, which charge a car battery, which in turn powers a sophisticated amateur transceiver, were in use long before the energy shortage "hit home" in the late 1970s.

virtually all types of voice transceivers—AM, FM, and SSB—and it will be particularly applicable in the VHF and UHF frequencies.

A great deal of applications development still remains to be done with NBVM and, as always, radio amateurs will play an important role.

The Challenge of the Unknown

What is still on the horizon in amateur radio? How about bouncing signals off the planet Venus? One amateur, who finds moonbounce "old hat," is exploring the radio astronomy techniques and technology involved in reaching Venus with a radio signal. He believes that the necessary expertise already exists in the amateur ranks and that successful detection of an echo from Venus is a distinct possibility in the not-too-distant future!

Want to Learn More?

If this book has gotten you motivated toward pursuing amateur radio further, it has served its purpose. First step is to write the American Radio Relay League, Newington, Connecticut 06111, for free information. They will put you in touch with a licensing class in your area and will tell you how you can join the only nationwide organization of hams in the United States. Aside from adding your voice to the 165,000 other hams and future hams who belong to ARRL, you will receive the League's monthly journal, *QST*, which is packed with articles and columns that will keep you abreast of the latest developments in the ever-growing field of amateur radio. The League's extensive list of books and pamphlets covers all aspects of the hobby, from setting up a ground station to listen to OSCAR satellites to learning the Morse code in preparation for the Novice exam. With any luck at all, you'll be operating your own amateur radio station in no time at all, and the means of making friends worldwide will be at your fingertips.

Glossary

AM	Amplitude modulation—voice transmission
ARRL	American Radio Relay League—national association of amateur radio operators
ATV	Amateur television—fast scan, as contrasted with SSTV, or slow scan
Band	A specific range of radio frequencies
Beam	A directional antenna popular with advanced amateurs
Coax	Coaxial cable, often used to connect the transmitter to the antenna
Contest	Making as many contacts as possible under specified conditions in a specified period of time
CQ	Standard amateur call on the air when seeking anyone to establish contact with

CW .... Continuous wave—Morse code transmission

Dipole .... A popular type of amateur antenna, especially with beginners

DX .... Distance—contact with amateurs in other countries, especially remote or uncommon locations

DXCC .... Famous ARRL award for verified contact with amateurs in 100 or more countries

Elmer .... An amateur who helps newcomers get into ham radio

FAX .... Facsimile—transmission of still pictures by radio

FCC .... Federal Communications Commission, the U.S. government agency that issues amateur radio licenses and regulates the service in the United States

Field Day .... A popular ARRL operating event held annually on a June weekend

FM .... Frequency modulation, a type of radio-wave transmission

Frequency .... Number of cycles per second of a radio wave

Gear .... Amateur radio equipment

Ham .... An amateur radio operator—the term is of uncertain origin but probably derives from the early telegraph days

Hamfest .... An informal but scheduled amateur radio get-together

HF .... High frequency—the amateur bands from 80 through 10 meters

Homebrew .... Slang term for amateur equipment built by a ham

Landline .... Telephone

MARS .... Military Affiliate Radio System, a communications service administered by the military and manned by radio amateurs

Meter .... A unit of measure for radio bands

Mobile .... Amateur radio operation in cars or other vehicles

Mode ....The type of amateur transmission—voice, code, TV, and so on

Net...................A group of amateurs who meet on the air on a regular schedule—usually to pass messages

Ohm's law ....Basic electrical ratio—voltage = current/resistance

OM ....Old man—traditional ham term of address for male amateurs

OSCAR ....Orbiting Satellite Carrying Amateur Radio—a series of communications satellites designed and built by amateurs and launched by NASA

Phone................Voice transmission

Pileup ....A large number of hams trying simultaneously to make contact with one station, usually a rare DX station

Q-signals............A universal system of amateur radio abbreviations understood in all languages

RACES ....Radio Amateur Civil Emergency Service, an emergency communications service

Rag chew ....An on-the-air conversation

Receiver ....A piece of equipment that receives radio signals and converts them to audible signals

Repeater............A device that receives and retransmits radio signals, extending their range

RFI/TVI.............Radio-frequency interference—occurs when unwanted signals are received by television receivers, stereos, and other equipment

Rig...................Ham term for a station

RST..................A system for reporting the quality of a radio signal to the sender

RTTY ....Radioteletype

Shack ....The room in which a station is located

Skip ....The propagation of radio waves that makes long-distance communication possible

SSB ....Single sideband—a type of voice transmission

SSTV ....Slow-scan television
Third party ....Anyone other than the operator for whom a message is sent via amateur radio
Ticket ....Ham slang for an amateur license
Traffic handling ....Relaying messages via amateur radio
Transceiver ....A unit that combines a transmitter and a receiver in one cabinet
Transmitter ....A piece of radio equipment that converts an audio signal into a radio frequency emission
UTC ....Universal Coordinated Time—time at a reference meridian used so that time designations are the same at every point in the world

Index

B

C

D

E

R

S

T

U

V

W